DISTRIBUTION OF REFINERY PRODUCTS IN INDIA

DISTRIBUTION OF REFINERY PRODUCTS IN INDIA

R. JAYASHANKAR

CONCEPT PUBLISHING COMPANY,
NEW DELHI

The Indian Council of Social Science Research gave the financial support towards the publication of this work. However, the responsibility for the contents of the book—facts, opinions, conclusions, etc., is entirely that of the author and not of the ICSSR.

First Published 1987

R. Jayashankar (*b*. 1954-)

Published by

Naurang Rai
Concept Publishing Company
H-13, Bali Nagar
NEW DELHI-110015 (India)

Printed in India by

M.J.S. Printers, Hari Nagar,
at Kiran Mudran Kendra, Mayapuri
NEW DELHI-110064

WORDS OF THANKS

I am extremely grateful to Professor A.S. Narang of the Faculty of Management Studies, University of Delhi who provided the intangibles of successful supervision. I am indebted to Mr. P. K. Agarwal, Joint Director, Systems, Oil Coordination Committee, New Delhi. I have benefited immensely from several educative, informative and purposeful discussions with him and others in the Oil Coordination Committee. The cooperation extended by the staff of the Oil Coordination Committee, Indian Oil Corporation Ltd., Bharat Petroleum Corporation Ltd., Hindustan Petroleum Corporation Ltd., Department of Petroleum under the Ministry of Energy, Indian Institute of Petroleum, National Productivity Council and Petroleum Conservation Research Association is gratefully acknowledged. I am unable to name them individually, for if I begin doing so, I do not know where to begin and where to end.

The models were evaluated on IBM 360/44 at the Computer Centre, University of Delhi. I am thankful to the Centre for permitting to me to have free access to the Computer. The financial assistance rendered by the University Grants Commission during the course of this research work is gratefully acknowledged.

R. JAYASHANKAR

CONTENTS

1

INTRODUCTION

The technological revolution of the twentieth century has been centered on crude oil. It is difficult to detect any facet of modern life where petroleum and its derivatives do not play a vital role. Nor can we forsee a situation in the near future when man can do without this liquid gold.

1.1 Refinery Products

Petroleum industry is characterized by the fact that all the refinery products are obtained simultaneously from one raw material, namely, crude oil. The various products can be grouped under the following heads :

(*I*) *Light distillates*

(a) Naphtha (b) Motor spirit (c) Liquified petroleum gas.

(*II*) *Middle distillates*

(a) Superior kerosene oil (b) Aviation turbine fuel (c) High speed diesel (d) Light diesel oil.

(*III*) *Heavy distillates*

(a) Bitumen (b) Fuel oil and a host of other products.

1.2 Consumption of Refinery Products in India

Consumption of refinery products in India has been growing quite rapidly in the last two decades. Table-1.I gives the consumption pattern of light, middle and heavy distillate products from the years 1970 to 1984. Each refinery product is consumed in one or more of the major sectors of the economy—industry, transport, agriculture, power and

TABLE 1.I : **Consumption of Refinery Products**

(Figure in '000' Tonnes)

Product	1970-71	1975-76	1979-80	1980-81	1981-82	1982-83	1983-84*
TOTAL	17912	22447	29883	30896	32523	34657	35488
1. Light Distillates of which ;	2697	3596	4460	4388	5187	5420	5541
LPG	176	336	410	404	492	601	744
Motor Spirit	1453	1275	1490	1522	1599	1722	1882
Naphtha	904	1836	2413	2325	2963	2958	2780
2. Middle Distillates	9040	11563	16321	17056	17919	19660	20455
Superior Kerosene Oil	3283	3104	3872	4228	4693	5214	5529
Aviation Turbine Fuel	689	897	1144	1125	1128	1145	1196
High Speed Diesel	3837	6595	9801	10345	10832	12013	12441
Light Diesel Oil	1092	878	1266	1122	1036	1067	1082
3. Heavy Ends of which ;	6175	7198	9102	9452	9417	9577	9432
Furnace Oil	3513	4279	4796	5406	4306	4208	4157
Bitumen	777	690	1069	1064	1292	1379	1028
Lubes/Greases	545	441	566	593	592	604	605
Petroleum Coke	107	151	185	137	174	103	112
Others	82	135	201	185	175	190	203

* Provisional

Source : Indian Petroleum & Petro-Chemical Statistics 1983-84, Economics & Statistics Division, Department of Petroleum, Ministry of Energy, Govt. of India.

domestic. Changes in the consumption of individual refinery products are determined by changes in the activity levels of the consuming sectors as well as the efficiency of utilization of these products. The sector-wise details of consumption of selected individual products for the years 1980-81 and 1981-82 are given in Table-1.II. It may be noted that industry and transport are the two main consumer sectors for the refinery products. Naphtha and fuel oils are the major products consumed by the industrial sector while motor spirit and diesel are consumed in the transport sector. The domestic sector consumed kerosene oil and liquified petroleum gas. Agriculture consumes high speed diesel oil in tractors and light diesel oil in irrigation pumps.

TABLE 1.II : **Estimated Sectorwise Consumption of Selected Petroleum Products**

(Figure in '000' Tonnes)

	1980-81		1981-82	
	Consumption	% to Total	Consumption	% to Total
1	2	3	4	5
Naphtha				
1. Fertilizers	1718	73.9	2316	78.2
2. Petro-Chemicals	595	25.6	639	21.5
3. Others	12	0.5	8	0.3
Total	2325	100.0	2963	100.0
Liquified Petroleum Gas				
1. Domestic	327	80.7	424	86.2
2. Commercial	20	4.9	16	3.2
3. Industry	57	14.1	51	10.4
4. Others	1	0.3	1	0.2
Total	405	100.0	392	100.0

(Contd.)

Table 1.II—(Contd.)

1	2	3	4	5
High Speed Diesel				
1. *Transport*				
(a) Railways	759	7.3	902	8.3
(b) Waterways :				
(i) Coastal & Inland	43	0.4	61	0.6
(ii) International Bunkers	15	0.2	17	0.2
(c) Other Transport	8233	79.6	8525	78.7
Total Transport	9050	87.5	9505	87.8
2. Agriculture	101	1.0	98	0.9
3. Power Generation	194	1.9	177	1.6
4. Iron & Steel	47	0.5	50	0.5
5. Textiles & Fibre	165	1.6	151	1.4
6. Cement	23	0.2	25	0.2
7. Chemicals & Allied	64	0.6	53	0.3
8. Aluminium	3	—	3	—
9. Sugar	30	0.3	33	0.3
10. Mining	184	1.8	214	2.0
11. Engineering	85	0.8	86	0.8
12. Fertilizers	13	0.1	31	0.3
13. Others	382	3.7	401	3.7
Total	10345	100.0	10832	100.0
Light Diesel Oil				
1. Transport				
(a) Railways	6	0.5	6	0.6
(b) Waterways				
(i) Coastal & Inland Bunkers	32	2.9	37	3.6
(ii) International Bunkers	51	4.5	52	5.0
(iii) Other Transports	1	0.1	2	0.2
Total Transport	90	8.0	97	9.4

(Contd.)

Table 1.II—(Contd.)

1	2	3	4	5
2. Agriculture	38	3.4	30	3.9
3. Power Generation	158	14.1	178	17.2
4. Iron & Steel	15	1.3	14	1.4
5. Cement	2	0.2	3	0.3
6. Chemicals	54	4.8	50	4.8
7. Aluminium	1	0.1	1	0.1
8. Sugar	4	0.4	4	0.4
9. Mining	4	0.4	5	0.5
10. Engineering	56	5.0	72	6.9
Total	1122	100.0	1036	100.0
Furnace Oil				
1. Transport				
(a) Railways	54	1.0	75	1.8
(b) Waterways :				
(i) Coastal & Inland	129	2.4	150	3.5
(ii) International Bunkers	149	2.7	67	1.6
(c) Other Transport	9	0.2	24	0.5
2. Agriculture	214	3.9	159	3.7
3. Power Generation	903	16.7	629	14.6
4. Iron & Steel	366	6.8	235	5.5
5. Textiles	471	8.7	342	7.9
6. Cement	166	3.1	51	1.2
7. Chemicals	780	14.1	673	15.6
8. Aluminium	113	2.1	98	2.3
9. Mining	275	5.1	131	3.0
10. Engineering	195	3.6	218	5.1
11. Fertilizers	585	10.8	617	14.3
12. Others	756	14.0	659	15.4
Total	5406	100	4306	100

Source : Indian Petroleum & Petro-Chemicals Statistics, 1982-83, Economics & Statistics Division, Department of Petroleum, Ministry of Energy, Govt. of India.

1.3 The Refineries in India

Until the early 1950s, India was almost entirely dependent upon imports for requirements of all kinds of petroleum products. At present the country has the following refineries :

(i) The Digboi refinery, Digboi
(ii) Bharat Petroleum Corporation Ltd., Bombay
(iii) Hindustan Petroleum Corporation Ltd., Bombay and Vizag
(iv) Indian Oil Corporation Ltd :
  (a) Barauni Refinery, Barauni, Bihar
  (b) Gauhati Refinery, Gauhati, Assam
  (c) Haldia Refinery, Haldia
  (d) Koyali Refinery, Gujarat
  (e) Mathura Refinery, Mathura, Uttar Pradesh
(v) Cochin Refinery Ltd., Cochin
(vi) Madras Refineries, Manali, Madras, and
(vii) Bongaigaon Refinery and Petrochemicals Ltd., Assam.

The above refineries can be classified into three groups according to their location : firstly the inland refineries, situated near the source of indigenous crude oil, such as the refineries at Digboi, Gauhati, Bongaigaon, and Gujarat; secondly the inland refinery based on domestic crude but situated near the demand centres. The Barauni refinery which processes crude oil transported from Assam oil fields is an example of this type of location; thirdly coastal refineries based on imported crude, such as the refineries of Bombay, Vizag, Cochin and Madras. The location of five coastal refineries based on imported crude combines the advantages both of resource-oriented and market-oriented refineries since these ports are also important areas of consumption. The possibilities of transporting surplus products by coastal tanker have contributed to the increase in the size of refineries at these port locations. Table-1.III summarizes the petroleum refining capacity in India as on 1.4.1983.

1.4 Problems of the Petroleum Industry

The many facets of the petroleum industry viz. (a) exploration, drilling, production and refining of crude oil and

TABLE 1.III : **Petroleum Refining Capacity in India**

(*Figure in Million Tonnes Per Annum*)

Refinery	Year of Commissioning	Capacity as on 1.4.1983	Capacity after expansion under implementation	Likely date of Completion
A. Refineries in Operation				
1. IOC, Digboi	1901	0.50		
2. BPCL, Bombay	1955	5.25		
3. HPCL, Bombay	1954	3.50		
4. HPCL, Vizag	1957	1.50		
5. IOC, Gauhati	1962	0.85		
6. IOC, Barauni	1964	3.30		
7. IOC, Koyali	1965	7.30		
8. IOC, Haldia	1974	2.50		
9. IOC, Mathura	1982	6.00		
10. CRL, Cochin	1966	3.30		
11. MRL, Madras	1969	2.80		
12. BRPL Assam	1979	1.00		
	Total	37.80		
B. Refineries under Expansion				
1. BPCL Bombay		5.25	6.00	October, 1984
2. HPCL, Vizag		1.50	4.50	October, 1984
3. MRL, Madras		2.80	5.60	July, 1984
4. CRL, Cochin		3.30	4.50	June, 1984

Source : Indian Petroleum & Petro-Chemical Statistics, 1982-83, Economics & Statistics Division, Department of Petroleum, Ministry of Energy, Govt. of India.

(b) distribution of refinery products, present a range of problems.

(i) The problems relating to petroleum exploration decisions under conditions of risk and uncertainty have been with us since the oil business began. If the existence of an oil reservoir has been established by the drilling of a productive well, problems arise in determining the order of magnitude of the hydrocarbon-bearing structural volume, the associated areal extent of the reservoir. These factors determine the spacing and location of wells to be drilled for optimum economic exploitation.

(ii) The technical flexibility of modern refining process permits the adoptation of the product mix to fairly wide fluctuations of market conditions, but the system has tended to become less responsive with its growing complexities and size. The problem becomes even more complex when account has to be taken of the economic variables in addition to the technical ones that is changes in market demand and prices both for refinery inputs and outputs. Thus given a government policy of achieving and maintaining self-sufficiency in refinery products, the domestic refineries are likely to produce surplus in some products and deficits in others, unless the overall supply of all refinery products equals the overall demand. Any shortage situation puts immense burden on the management of the distribution sector. The pattern of refinery production is summarized in Table-1.IV.

(iii) Distribution covers a range of physical operations and includes the transportation, storage, handling and transmission of refinery products to the points of consumption. Reliability and regularity of supply are of vital importance to any economy. For example, the demand for high speed diesel has been growing quite rapidly during the last few years. Pointing out the poor performance of the industry on account of meagre power and diesel availability, the association of Indian Engineering Industry has recently sent a memorandum to the Department of Petroleum under the Ministry of Energy, to review the diesel distribution system. Shortage of furnace oil upsets industry schedules. Farmers are plagued by shortage of light diesel oil, who need the product to energise their field

TABLE 1.IV : **Pattern of Refinery Production**

(Figure in Percentages)

Products	1979-80	1980-81	1981-82	1982-83
Light Distillates*	16.2	15.2	17.0	16.0
Middle Distillates	47 6	46.9	46.9	47.1
Heavy Fuel Oils	23.1	23.7	22.0	24.0
Bitumens	4.0	4.2	4.3	4.2
Lubes	1.8	1.7	1.4	1.3
Others	1.2	1.0	1.0	1.1
Total Production	93.9	93.4	93.5	93.7
RBF/Losses/Inventories	6.1	6.6	6.5	6.3
Crude Run	100	100	100	100
*Includes Gases	1.5	1.4	—	—

Source : Indian Petroleum and Petro-Chemicals Statistics 1982-83, Economics & Statistics Division, Department of Petroleum, Ministry of Energy, Govt of India.

pumps. Thus building up a distribution network is a matter of continual balancing of alternatives and searching for the least cost methods.

1.5 Functions of marketing oil Companies

The distribution functions of refinery products in India are performed by Indian Oil Corporation Ltd., Bharat Petroleum Corporation Ltd., and Hindustan Petroleum Corporation Ltd. The primary objectives of the marketing oil companies as pointed out by the Damle Committee Report on distribution system of petroleum products (1976) are as follows :

(i) From the point of view of the consumers, to provide refinery products of the prescribed specifications in accordance with the varying requirements of the consumers; in correct measure and requisite quantity; in suitable packages; as near to the consumers as marketing economics would allow; and at the prescribed selling prices.

(ii) From the point of view of the total cost to the economy, which inevitably get reflected in the retail/ceiling prices of the refinery products, to cut down distribution operating and marketing losses at all stages, i.e. at the storage points, during transportation by various means and at the delivery stage. While some losses in the course of storage and handling are unavoidable for volatile products such as motor spirit, superior kerosene oil, there is always scope for minimising these.

(iii) From the point of view of the national economy, under the changed conditions, where due to the manifold rise in oil prices, it has become necessary to exercise maximum possible economy in the consumption of refinery products, to educate the users in the proper application/consumption of products in the most efficient manner.

(iv) As organisation directly controlled by the Central Government, to set up standards of service to the consumers; and to act as model distributing agencies in all respects, including for regulating supplies of products that may be or may become deficit for varying durations in the most equitable manner and in the closest possible coordination with the State Governments.

1.6 Distribution Arrangements—An Outline

It would be useful to record in very broad terms the existing distribution arrangements, before a discussion of the plan of the present study. The product requirements are met by production from refineries and supplemented by imports. The refineries and the main port installations form the focal point of the countrywide distribution network. The upcountry depots from secondary distribution centres receiving supplies from one or more of the refineries and/or main port installations, predominantly by rail, but also to a limited extent by road and in some locations through product pipelines. The storage points at the last referred locations are known as pipeline tap-off points. As far as the general public is concerned, the last link in the chain of distribution is the retail outlet, popularly known as the pump. The retail outlet/pump receives the refinery products from upcountry depots. The major transport facilities for distributing the refinery products are the rail, road, coastal movement and pipelines.

1.7 Objectives of the Study

This study is divided into three main parts to facilitate the meaningful illustration of the distribution process from refineries, main port installations and upcountry depots. Also the study enables us to arrive at useful solutions to the major problems associated with the same.

(i) The first part deals with the distribution of a product from refineries. The heavy distillate product, bitumen is chosen for this purpose. The optimal production and distribution pattern from refineries to various states/union territories in bulk and packed form of the product is arrived at.

(ii) In the second part of the study distribution of light, middle and heavy distillate products among the main port installations by coastal tankers are considered and the optimal port traffic arrived at.

(iii) The third part is an indepth study of queueing problem of tank lorries in upcountry depots. For this study of the distribution from depots to retail outlets, high speed diesel is chosen, because the queue is longer for this product as compared to other refinery products.

The primary objective of any distribution arrangement is to move products from source to the ultimate consumer, at the most economic cost. To this end, the above studies can be conceived as a means for optimal distribution of refinery products at the least total cost to the national economy.

1.8 Scope

The distribution process from the focal points, namely, refinery and port and the secondary distribution centre, that is, upcountry depot, have been covered. However distribution from pipeline tap-off point and retail outlet/pump have not come under the purview of the study.

1.9 Methodology

For the study, primary data were collected from refineries, the Oil Coordination Committee and the oil companies. Secondly data was collected from the Department of

Petroleum, Ministry of Energy. Material was also studied, distilled and drawn in condensed form from various reports and statistics published by the Department of Petroleum and Oil Coordination Committee.

1.10 Approach

The approach in this study has been to develop mathematical models based on operations research tools like linear programming, goal programming, transportation techniques and queueing theory. With the aid of these models, simultaneous and synergetic decision making is achieved, i.e. the influences of all economic factors are simultaneously considered to arrive at optimal distribution patterns.

1.11 Limitations

The study is basically an analytical one covering various refinery products which are produced and consumed in a widely dispersed geographical area in the country. Besides the limitations, applicable to the operations research techniques, the data on demand and supply of refinery products released by said agencies are also subject to change. This is likely to serve as a statistical limitation to the study.

1.12 Plan of the Book

Chapter II is a survey of the literature pertaining to specific Indian studies and usage of operations research techniques for the petroleum industry. Chapter III gives a bird's eye view of the existing distribution system for the refinery products. Chapter IV deals with the distribution of a heavy distillate product from refineries. Chapter V focuses on the coastal movement of light, middle and heavy distillate products. Chapter VI is an indepth study of distribution of a middle distillate product from upcountry depots. Chapter VII is the concluding framework.

2

FINDINGS OF PREVIOUS STUDIES AND MAJOR ISSUES

This literature survey deals with specific Indian studies, related to the petroleum industry. The review of applications of operations research techniques for the petroleum industry is also presented in the view of the abundant usage of such techniques in this study. There is ample literature on the basic rudiments of the distribution process for refinery products [Sell, G, and H.A. Dossett (1958); Frankell, P.H. (1969); Hahn (1970); Marshal Sitting (1978)].

The earliest Indian study on marketing of refinery products was conducted by Foster Wheeler Corporation as a part of their study on petroleum refinery survey for the Government of India. The Foster Wheeler Corporation conducted two studies, one in 1958, and the other one in 1963. The 1958 study examined the freight rate structure of various modes of transportation for the refinery products. Also the average losses in the process of importation, storage and distribution for different products were discussed.

The 1963 study mainly dealt with the imbalance of supply and demand of refinery products. The relative consumption of motor spirit, high speed diesel, aviation turbine fuel and furnace oil during the sixties had undergone charges due partly to consumer's preferences and partly to technical economic and social factors. The study also concluded that the demand for middle distillate products will be more and that the refineries will produce surplus of some products and less of others. Das Gupta (1971) has also presented a detailed analysis of product imbalances in the Indian petroleum industry. He has shown that the following factors were responsible for the serious imbalances in consumption and production :

(a) The relative prices obtained by the refiners favoured output of light distillates rather than middle distillates;
(b) The relative prices for consumers encouraged the use of middle distillates;
(c) The estimation of demand for motor spirit and kerosene was faulty.

The Damel Committee on the distribution system of petroleum products was set up by the Government of India in 1975 to examine and report on the following:

(i) The arrangements for the distribution of the petroleum products starting from refinery down to the retail outlets.
(ii) The need to expand the retail network, so as to reach the products as close to the consumer, particularly in the rural areas.
(iii) Whether the retail network should be further strengthened by induction of genuine consumer cooperatives, the agro-industries cooperation, agro-service centres in the rural and semi-urban areas.
(iv) Level of inventory to be maintained at the various storage points below which stocks should not be allowed to deplete.
(v) The manner in which oil companies may be designated as regional lead companies for maintaining close liason with the State Governments and its agencies to ensure that adequate supplies are maintained and distribution of consumer products like kerosene, HSD and LDO is done on a rational basis.
(vi) The means necessary for economising on overheads, transportation, cross movement and other charges to reduce their incidence on consumer prices.

The Damle Committee submitted an interim report in September 1975 regarding necessary steps to meet the petroleum product needs of the agricultural sector. In the final report, for the first time, the primary objective of the marketing oil companies from the point of view of the consumer and the economy as a whole was defined. The Committee recommended regional demarcation of the marketing operations of the oil companies. The Committee suggested complete integration of the storage and distribution facilities of different oil companies at the upcountry locations. The Committee had also laid stress on developing machinery for correctly assessing the demand for petroleum products, and for clearly demarcating the functions of the oil companies, the Oil Coordination Committee and the Ministry of Petroleum. The Committee also stressed the need for developing a comprehensive computer model which would give the relative merits and costs of the various alternatives within the known constraints of determining the refinery-wise operating levels, the type of crude oils to be supplied to each refinery, the determination of production pattern and the total distribution costs. It may be mentioned that the Damle Committee Report is the only comprehensive study on distribution of refinery products in India.

The other reports namely, the Fuel Policy Committee Report (1974) and the Report of the Working Group on Energy Policy (1979) had not dealt much with the distribution of refinery products. On the basis of a detailed system study of the refinery products demand including the demand for non-energy uses and the refinery investment choices, the Fuel Policy Committee made a strong recommendation that a larger investment should be made in secondary processing like hydrocrackers, catalytic crackers which would convert the surplus heavy end products to middle distillates and thus minimise the supply-demand imbalance. The Working Group on Energy Policy (1979) stressed the need for less energy intensive modes of transportation for movement of goods including refinery products. Also the Report stressed the need for coordination of rail and road traffic for movement of refinery products.

Jyoti K. Parikh in a report entitled "Modelling Energy demand for policy analysis" (1981) has very rightly justified the due consideration to be given for spatial aspects or regional distribution of energy. In general, national level models on energy, do consider sectoral allocations of natural and financial resources. Thus the disaggregation of plan targets in terms of sectors, such as industries, agriculture and so on is available. However, breaking up of these targets into state level targets has been a difficult task partly because of the constitutional autonomy of the states and also due to the lack of methodology for optimally allocating the resources over a long term. Moreover, even if it is theoretically possible to determine such an optimal strategy, national level optimum may not be suitable to individual states desire for their own chosen paths of development. Since a bulk of financial and natural resources are contributed by the states themselves, it is necessary that allocation of plan targets would be acceptable to the states as well. Thus the importance to have regional distribution of demand for refinery products to reduce transport bottlenecks was highlighted in the study.

The National Transport Policy Committee (1980) had broader terms of reference. They were :

(a) To propose a comprehensive national transport policy for the country for the next decade or so keeping in view the objectives and priorities set out in the five-year plan.
(b) To identify the areas in which the data base of the transport system should be strengthened in order to be able to formulate integrated transport plans.
(c) To recommend areas in which research and development in the transport field should be undertaken and the institutional framework for carrying out the same.
(d) To suggest measures for improving training facilities in transport planning and management.

The Committee examined in detail the role played by rail, road, coastal movement, pipelines and measured the resource cost of each agency of transport on a comparable basis.

The Report of the Oil Prices Committee of the Government of India (1969, 1976) presented a discussion involved in pricing of refinery products. A pricing formula for determination of selling price at the main installation and upcountry depots was evolved. Accordingly the selling price at the main installation and upcountry depots should be on the basis of supply coming from the nearest source by the cheapest mode of transport.

Ketharaman, et al (1973) outlined a few important problems and practices relating to marketing and distribution of petroleum products mainly with reference to the experience in India. Salient features affecting the oil industry in India were also outlined.

Budhiraja, (1978) stressed the importance of reliable forecasting of demand and production planning as prerequisites for effective distribution of refinery products.

The petroleum industry became aware of operations research techniques through the pioneering work of Alan S. Manne (1956), studies by Manne illustrated the significance of the joint product nature of petroleum refining industry necessitating consideration of technical and economic interrelations which determine the product mix of the refineries. Alan S. Manne evolved a linear programming model of the United States petroleum refinery industry (1958). Earlier Charnes, Cooper and Mellon in the paper entitled "A model for programming and sensitivity analysis in an integrated oil company", studied the optimal assignment of crude oils between the various refineries of a multiplant firm (1954). Symonds, (1955) had shown to use linear programming to find the most profitable plan for processing crude oil in a hypothetical refinery.

Garvin, Crandoll John, and Spellman made a survey of the applications of linear programming by a number of oil companies (1957). Examples were presented of applications to a variety of problems arising in the areas of drilling and production, marketing and distribution. Examples were selected to illustrate both the power and the limitations of linear programming methods when applied to actual problems.

Cheney, (1957) of Richfield oil corporation showed how complex technical economic problems encountered throughout the entire petroleum industry may be effectively evaluated by use of linear programming techniques. Problems tackled included simple gasoline blending. catalytic reforming and finished product mix model. The matrix of the problem contained 66 variables and 27 constraints.

The optimum routing of a fleet of gasoline delivery trucks between an oil depot and a large number of retail outlets was studied by Dantzig and Ramser (1959). The shortest routes between any two points in system were given and a demand for one or several products was specified for a number of stations within the distribution system. It was desired to find a way to assign retail outlets to trucks in such a manner that station demands were satisfied and total mileage covered by the fleet was minimum. A detailed procedure based on a linear programming formulation was given for obtaining an optimal solution.

Newby and Dean (1960) have shown the way the British Petroleum Company made use of linear programming for the setting of its short and medium term operations. Sellars, and Strang (1962) emphasized on the optimization of the operations of a group of refineries and marketing activities by the use of linear programming and computer techniques.

Koeing (1971) examined the problems of long-term planning and day-to-day planning of refineries by linear programming technique. Linear programming formulation assumes that technical relations in refinery operations are linear equations. It was suggested by Koeing that about 95 per cent of all refinery functions can be expressed by linear equations, about 4 per cent are nonlinear but convex, and 1 per cent non-linear concave.

Balashova, et al in the paper presented at the VIII World Petroleum Conference (1971), dealt with the application of operation research techniques for future planning of the refineries, the system of oil industry as a whole, its subdivisions and connections with other branches of national economy were analysed. A performance criterion was chosen, so that account could be taken of the systems external relations. The paper concluded with the formulation of the stochastic problem of optimal planning and some approaches to its solutions.

Moore, and Zoltners examined the potential use of linear programming in developing an oil allocation plan (1971). The primary objective was to provide an economic model and a method of analysis which would help planners and policy-makers determine the most efficient distribution of refinery products among industries and regions in a nation. A second objective was to develop an economic model for simulating the impact of a particular allocation program on the economy at both the national and regional levels. This was highly desirable because of the critical impact such programs have no prices, employment, production and the basic quality of living.

The model as developed by Bhatia (1976) was a multiprocess interregional programming model for analyzing technological and locational choices in petroleum industry in India. The partial equilibrium model, assumed the prices of capital, labour, foreign exchange and discount rates as given with the objective of minimizing the total cost of meeting exogenously determined demand for petroleum products and nitrogenous fertilisers for a target year. The model results indicated optimum levels of (i) Foreign trade in crude oils, petroleum products, and fertilizers (ii) Process activity levels in refining such as simple refining, hydrocracking, coking etc. at each location and (iii) transportation of crude oils, intermediate products and final products.

The optimum locations for the bottling plants for liquified petroleum gas (LPG) in India and their suggested capacity have been worked out by Sumit Roy, Gautam Roy and Banerjee (1982). The integer programming model considered the following factors :

(a) Transportation cost of bulk LPG from source point to filling plants.
(b) Transportation cost of filled cylinder from filling plants to consumption centres.
(c) Filling cost of LPG.

The literature survey permits certain inferences.

(i) Data and studies relating to distribution of refinery products in India are not adequate.
(ii) Operations research techniques have immense potentialities for many intricate problems of the refineries.

The significance of the use of operations research techniques is adequately realised in India, and the Indian Petroleum Industry has resorted to the use of linear programming. However, such usage is restricted only to the production planning problems of the refineries. There is hardly any study on the applications of operations research techniques for distribution problems of refinery products. In the present study, apart from linear programming, other operations research techniques like goal programming and queueing theory have been artfully exploited, to provide an insight into some of the distribution problems of the refinery products in India.

3

THE EXISTING DISTRIBUTION SYSTEM

3.1 Organizational Set Up

The Ministry of Petroleum and Chemicals came into being in November 1963, with two departments, namely the Department of Petroleum and the Department of Chemicals and Fertilizers. At present the Department of Petroleum is under the Ministry of Energy.

(i) The Department of Petroleum is divided into the following divisions :

(a) Exploration division
(b) Marketing, distribution and coordination division
(c) Refineries division
(d) Pricing division
(e) Petrochemicals division
(f) Finance and Accounts division
(g) General Administrative division

(ii) The functions assigned to the Marketing, distribution and coordination division include assessment of requirement, availability and distribution in the country of petroleum products, allocation of kerosene/diesel to states/union territories, prevention of adulteration of petroleum products, demand forecasting, sales plan of oil companies, problems relating to scarcity of petroleum products, matters relating to retail outlets, dealership of kerosene, diesel and petroleum agencies, complaints in regard to malpractices curbs on consumption of petroleum products, fuel conservation and efficiency measures.

(iii) The following undertakings are controlled by the Marketing, distribution and coordination division of the Department of Petroleum.

1. Indian Oil Corporation Ltd. (Marketing division)
2. Bharat Petroleum Corporation Ltd. (Marketing division)
3. Hindustan Petroleum Corporation Ltd. (Marketing division)
4. Indo Burma Petroleum Company Ltd.

(iv) Other organisations under the administrative control of the Department of Petroleum include the oil industry development board, oil coordination committee and the petroleum conservation research association. The oil coordination committee coordinates the planning of supplies and demand projections for efficient distribution of petroleum products and works closely with the oil industries and the Department of Petroleum, Ministry of Energy.

The Oil Coordination Committee (OCC) was established by the Ministry of Petroleum in 1975 indicating the acceptance of the recommendations of the Oil Pricing Committee (OPC) in its interim report. It is necessary to comment in some detail on the importance of a coordinating machinery in the present context with special reference to the role that is usefully played by the Oil Coordination Committee in this regard. The refineries are not isolated production units, even where these have been established primarily to meet the regional demand. To achieve maximum overall efficiency, the allocation of crude must be so determined for the country as a whole, that the need for imports is minimized and the total distribution costs are optimised. With all imports now being regulated by the Government, it has become possible to determine which type of crude, each with its special characteristics, needs to be earmarked for various refineries in the interest of obtaining optimal refinery results, with preference to the needs of each producing area, loading facilities, minimization of movement costs and related considerations such as the balancing of surplus and deficits from these operations. All this involves not only a very careful and detailed overall planning of crude and product imports and refining programme but also a well calculated policy to ensure a healthy and much-needed build-up of stock during the slack season, particularly of products which are highly sensitive to seasonal variations in demand, so that there are minimum stresses and strains on the available movement capacity by various means. The plans thus made by the coordinating agency have also to be subjected to constant adjustment in the light of everchanging situations and developments based on the feedback obtained from the countrywide monitoring of the supply and demand position.

(i) *Functions of the OCC*

(a) *Supply planning activity* : The OCC coordinates the supply planning activity for the oil industry working closely with the industry and ministry. Activities pertaining to this function are as follows :

1. Preparation of the oil economy budget : This indicates the crude and product imports requirements and serves as the basis for the annual plans and financial approvals.
2. Monthly industry coordination meeting : In this regular meeting, the supply outlook for the next 3-4 months is reviewed and plans for additional imports discussed. The meeting also serves as a forum for discussions on other topics of interest. The participants include representatives from the oil industry.
3. Supply plan meetings : This meeting is conducted by the Department of Petroleum for which the necessary support is provided by the OCC Secretariat. The plans for distribution and movement of petroleum products during the forthcoming month are finalised at this meeting.
4. Monitoring activity : In coordination with the Department of Petroleum and oil companies, a constant follow-up is maintained on the actual operations during the month and wherever necessary, corrective actions are taken for ensuring supplies.
5. Crude Imports : The OCC coordinates upliftment and distribution of crude to refineries considering both the

crude quality requirements and the need for maintaining crude supplies for refinery operations.

(b) *Optimisation of industry operations*: This includes a constant review of operations and special studies for improvement. Activities pertaining to this function are as follows :

(1) Evaluation of crudes based on their yield patterns and prices.
(2) Optimal allocation of crudes to the various refineries.
(3) Optimisation of refinery production slates keeping in view the demand requirements.
(4) Studies on Bombay High absorption outlook.
(5) Studies on LPG absorption outlook.
(6) Studies on ocean losses.
(7) Review of product specifications.

(c) *Management of pool funds*: This includes the administration of pool accounts and related activities, namely

(1) Disbursement of claims and surrender by the oil companies to the pool accounts.
(2) Studies on projected outlook for pool fund balances.
(3) Studies for revision in pricing of petroleum products.
(4) Revision of refining and marketing costs, and retention prices.

3.2 The Distribution Network

(i) The petroleum product requirements are mostly met by the production from the refineries, supplemented by imports. Presently, the twelve operating refineries and the main port installations from the focal points of the countrywide distribution network. The upcountry depots form Secondary distribution centres receiving supplies from one or more of the refineries and/or main port installations, predominantly by rail but also to a limited extent by road, and in some locations through product pipelines. The storage points at the last referred locations are known as "pipeline tap-off points". Large customers and bulk customers like steel plants, railways fertilizer plants, industrial consumers, power houses and so on draw their supplies of product directly at their own railway sidings and storage facilities. Some of the larger consumers located in close proximity to refineries/main installations receive supplies directly by pipelines. Supplies of aviation produts form a distinctive operation, often with their own specialized arrangements. However, as far as the general public is concerned, the last link in the chain of distribution is the retail outlet, popularly known as the pump, except in the case of kerosene, which is a trade of small volumes purchased by numerous individual customers through a multiplicity of dealers and small traders including hawkers. Table 3.I shows the existing position in regard to operating refineries, main port installations, pipeline tap-off points, upcountry depots and retail outlets forming the vast distribution network in the country as own 1982-83.

TABLE 3.I : **Distribution Network as on 1982-83**

Operating refineries		
	Coastal	5
	Inland	7
	Total	12
Main port installations		11
Pipeline tap-off points		10
Upcountry depot locations		142
Retail outlets		11,884

Source : *Indian Petroleum & Petro-Chemical Statistics 1982-83*, Economics & Statistics Division, Department of Petroleum, Ministry of Energy.

(ii) The primary objective of any distribution arrangement is to move products from source to the ultimate consumer, at the most economic cost. This implies, among other things choosing the cheapest and the most suitable mode of transport from the alternatives available, reducing to the extent possible the number of stages in the chain of distribution, handling the largest possible volumes at each stage and securing the most intensive use possible of the available resources. The current pricing formula for petroleum products lays down that prices at the main installations and upcountry depots should be determined on the basis of the prices at the nearest refinery point plus the cost of transport by the cheapest mode—coastal, rail, pipeline or road movements. Such costs for price computations are necessarily linked to the oil industry itself but in the final analysis, costs have to be viewed in the larger context of 'Social, and 'national' costs. The underlining principle of the current pricing formula is thus the determination of the selling price at any given location, on the basis of supply coming from the nearest source, by the cheapest mode of transport. This is a sound principle and unexpectionable by any standards. This also provides the scope for minimisation of costs due to the fact that in many cases, the supply points do not have products in exactly matching quantities to meet the full requirements of their economic pricing areas. Some have more than what can be absorbed in the concerned pricing area, necessitating product transfer at additional costs to other pricing areas, while others have deficits which need to be made good from the surplus supply points. The minimisation of surpluses and deficits and their management at the least total cost to the national economy, provide tremendous challenges and opportunities for employing the most sophisticated techniques and to bring about continuing refinements on an every changing scene, for it must be accepted that irrespective of how the pricing system is built up, the total costs would invariably have to be absorbed by the ultimate consumer of the oil products.

(iii) Against the background of these broad principles, certain features of the present distribution arrangements through rail, road, coastal tankers and pipelines are considered.

(a) *Rail* : The inland movement of petroleum products in the country is predominantly by rail. Some years ago, the adequacy of rail tank wagon availability for petroleum loading used to present several problems, necessitating imposition of 'no tank wagons' and road bridging of products at uneconomic cost, often to locations upto a distance of 250 kilometers or more, and generally imposing a severe strain on the distribution system. This situation in this regard has since undergone a major change. Tankwagon availability is now reported to have outpaced the demand. While there are several important reasons for this improvement, conscious efforts have been made in the past for debottlenecking of constraints on the side of both the railways and the oil industry. One strikingly contributing factor has been progressive strengthening of facilities at an increasing number of depot locations to receive full train loads of tankwagons. It has thus been possible to substantially reorganise the rail movement pattern of petroleum products more and more in block train loads which by pass the intermediate sorting rail yards, instead of piece-meal movements by attaching a few wagons to a normal freight train. The overall national cost advantages of sending products in full train loads are applicable for most commodities but are particularly pronounced in the case of oil products. This is because, unlike other general purpose wagons, which can be moved to any desired loading points and can also be back loaded with a large variety of freight traffic, tank wagons are special purpose wagons, allocated to specified loading bases and would in the normal course return empty to such bases, unless changes in the loading pattern necessitate the restencilling of the base marks on individual tank wagons for diversion to some other base. This system confers large national benefits to the transporter, in this case, the Indian Railway and also to the sender/receiver of the refinery products. Besides minimising investment on tank wagons, which cannot be readily switched over to the other use, it sharply cuts down transit time, enables a closer and more effective monitoring of replinishments and facilities a much better management of the planned loading and unloading schedules. Also investment on creation of storage capacity at the upcountry depots is optimized. The saving in storage capacity is achieved by relating the productwise storage capacity requirements to what can be transported in full train loads plus a suitable cushion, instead of following the conventional practice of relating the storage capacity to a certain number of days cover.

Subsequent increases in traffic to be handled could then be largely met by increasing the frequency of the block specials rather than by adding to the tankage to correspond to the progressively increasing offtakes. An incidental benefit of running such block train loads is that it reduces to a large extent, the risk element involved in the sorting of tank wagons in railway yards, particularly the very big one. Tank wagons cannot be loose shunted and have to be very carefully detached from one train, sorted out for various trains under formation at various lines, and then have to be attached to each one of these in accordance with the direction/destination of the reformed trains. The system of block train load movements has not, however, been carried to its logical conclusion. This is because the facilities for receiving block train loads at depots have been set up only by the Indian Oil Corporation which has invested heavily on matching storages, decanting headers and sumps, increased pumping capacity etc. As the other oil companies have not been able to augment their facilities to suit block train operations, due to their self imposed restraints on capital expenditure, the block train loads need to be split up at the receiving end for company wise sorting and placements. The tank wagon unloading sidings, in most cases specially constructed by the railways to accommodate block loads, cannot therefore be utilised to deal with the traffic of all the oil companies from the same common siding. This resutls in wasteful operations which, besides taking up valuable space in the railway yards, also add to the total transport costs due to increased detentions to tank wagons and additional shunting costs. From the side of the oil companies also, it is clearly necessary to rationalise their distribution activities by making use of each other facilities in the depots, thereby increasing the utilisation of fixed assets and reducing unit costs. Joint optimisation of depot costs by joint use of railway sidings, common loading and unloading arrangements and even joint use of storage facilities is a vital necessity. The economies achieved will be substantial both in terms of current costs and planned additions to storage facilities and would more than justify any efforts required to achieve this objective.

(h) Road movements : In tank truck operations, the ideal to be aimed at is direct delivery from the storage tanks of the oil companies to the storage of the customer or the retail outlet, by a fully loaded tank lorry of the largest convenient size and with a turnaround time which eliminates idling at the loading and/or the unloading points. There are however several limitations in actual practice.

(1) With the sharp increase in the price of refinery products in the last few years, retail outlet dealers now mostly look for part loads of Motor spirit. This happens even where the actual size of the road vehicle gets reduced by road conditions and axle load limitations imposed by bridges and culverts.

(2) As for the turnaround time of road vehicles, at most truck loading points, long queues of tank trucks waiting for loads in the early part of the day, is a common sight. At some of the locations queues are so long and the waiting time in many cases vary from three to five hours.

(3) There is thus considerable scope to economise on transportation costs as far as tank truck operations are concerned. One cannot look merely at the size of the direct savings in this regard, for savings would also be in the form of reduced fuel consumption of the most important middle distillate product viz. HSD, whose requirements almost directly determine the refinery capacity and/or import requirements of the country. The progressive elimination of these constraints, which have an important bearing on distribution costs, would require concerted action and long range planning.

(c) Coastal movement : The benefits anticipated from the elimination of foreign flag vessels and their substitution by Indian flag vessels have not been fully realised. Instead of there being a reduction in coastal movement costs, these have sharply increased, with consequent increases in the burden to the oil industry and, to the consumer at large. Break downs, sometimes threatening product dry outs and necessitating costly last minute diversions in planned movements are not uncommon. A detailed treatment of coastal movements is presented in Chapter V.

(d) Pipeline transport is ideally suited for refinery products and is conducive to substantial economies where there is a heavy concentration of traffic flow. Operating costs are comparatively less, maintenance relatively inexpensive and the movement is continuous. There is no problem of return of empty containers to loading points as in other modes of transport. Transit losses which are innerent in other modes of transport are considerably reduced. Yet despite its overwhelming advantages, pipeline transport has remained completely static in the last decade and a half and there has been no expansion of the system, notwithstanding all the increases in the overall consumption and movement levels that have taken place during this period. Table III shows the pipeline positions as on 1-4-1983. In this context, inter alia, the more important considerations that need to be highlighted are the following :

(1) The lowest per unit transport costs are incurred when refinery products are transported by pipelines.
(2) Considering the very large foreign exchange component involved in the import of crude oil/refined products, even from the point of view of transit losses, other modes of transport should be preferred only when low volumes preclude the need for making investments on pipelines.
(3) The railways are already highly over capitalised. It is true that certain expansions of the rail transport capacity, primarily for carrying passengers and goods, often result in surplus capacity becoming available for the transport of refinery products as well. Such a situation needs to be viewed, not only in the context of the points mentioned above, but also in an overall perspective of the anticipated traffic load on the railways in the next 15 to 20 years. If this is done, the overall national economic interest may justify the laying of the pipeline.

Table 3.II summarises the movement of petroleum traffic in Railways for years 1972 to 1982. Table 3.III gives the product pipelines position as on 1-4-1983.

3.3 Demand Forecasting of Refinery Products

(i) The first stage at which shortages are normally introduced in the national economy, is the area of forecasting of demands of commodities. In view of the resource constraints, there is always a tendency to estimate the demand pessimistically, and to forecast production in undue optimism. As reliable demand forecasting is the first prerequisite for an effective distribution system, a discussion of the same is presented in the following :

TABLE 3.II : **Movement of Pol Traffic on Railways Since 1970-71**

Year	Tonnes Originating (000' MT)	Earnings (000' Rs.)	Average lead (kms)
1970-71	8857	464081	594
1971-72	10063	523049	593
1972-73	10181	550982	614
1973-74	9994	555560	638
1974-75	10751	723192	642
1975-76	11663	860925	605
1976-77	12395	1038462	609
1977-78	13062	1120693	631
1978-79	14301	1324332	698
1979-80	14266	1497510	729
1980-81	14954	1835693	780
1981-82	16555	2848050	731

Source : Indian Petroleum & Petro-Chemicals Statistics, 1982-83, Economics & Statistics Division, Department of Petroleum, Ministry of Energy, Govt. of India, New Delhi.

TABLE 3.III : **Product Pipelines in India as on 1.4.1983**

Pipeline	Owned by	Length in kms.	Diameter in Inches
Product Pipelines			
Digboi Refinery—Tinsukia	IOC	34	6″
Digboi Refinery—Tinsukia	IOC	39	6″
Digboi Refinery—Tinsukia	IOC	34	4″
Gauhati—Siliguri	IOC	425.85	8-5/8″
Barauni—Kanpur	IOC	668.78	12-3/4″
Koyali—Ahmedabad	IOC	115.70	8-5/8′
Haldia-Mourigam-Rajbandh	IOC	268.58	12-3/4″
Mathura-Delhi	IOC	147.0	16″
Delhi-Ambala	IOC	202	14″
Ambala—Jullundur	IOC	164	12-3/4″
Total		2098.91	

Source : Indian Petroleum & Petro-Chemicals Statistics 1982-83, Economics & Statistics Division, Department of Petroleum, Ministry of Energy, Government of India, New Delhi.

(ii) The forecasting of demand has always been a hazardous exercise and is much more so when it has to be undertaken on perspective or long term basis. During the last few years even the preparation of annual forecasts of demand has become an exceedingly difficult task not only in India but in almost all the countries, including those which have readily available data with specific reference to the end use of each product. In the past this difficult and complex process has been attempted by bringing together the professional economist, the technocrat, planner as well as the experts of the marketing oil companies. In the final analysis the decisions have been taken by the Government itself as part of the overall national planning process, taking into account the broad relationships between the projected economic growth and energy requirements, the extent to which energy intensity (i.e. energy input per unit of product) can be obtained by more efficient usage and the total energy demand.

(iii) In the past the Indian Institute of Petroleum (IIP) periodically made studies in this regard. Some of the oil companies made their own independent demand projections. All these formed the subject matter of discussion in the Oil Advisory Committee before the Government took final decisions. The Oil Advisory Committee has since ceased to function and in order to coordinate the activities of the various agencies concerned with the preparation of demand forecasts of petroleum products and to ensure that the knowledge, experience and resources available in the country are effectively pooled, the Government constituted a "Demand Forecasting Cell" (DFC) in July 1971. This cell comprised the representatives of the Planning Commission, Finance Ministry, Railway Board, Indian Institute of Petroleum, Marketing Oil Companies, and a representative of the Ministry of Petroleum, who also acted as the Convenor of the Demand Forecasting Cell.

(iv) The broad objectives of the cell included, inter alia, the following :

(a) Short term, medium, long range and perspective forecastes of demand productwise, regionwise and sectorwise.

(b) Analysis of variations in demands vis-a-vis anticipation productwise, regionwise and sectorwise.

(c) Division of the country into a set of regions and sectoral classification of industry that could be of help primarily in forecasting demand.

(d) Working out surpluses and deficits based on demand forecasts and estimates of production on a short, medium, long range and perspective basis, for supply plan, trade plan and refinery capacity planning.

(e) The DFC using techniques like regression method, end use method, trend method etc., prepared reports covering the period from 1974 upto 1980 and also on a perspective basis upto 1984 i.e. for a total time span of 10 years. Unfortunately with the upheavals and the uncertainities caused by the unprecedented price increase in the late 1973, this organisation became dormant. As demand forecasting is a basic necessity for the efficient functioning of any distribution system, and in fact for the oil industry itself, the Demand forecasting cell was reactivated.

(f) In the Oil Coordination Committee (OCC) an industry committee was set up for deciding on the allocation of crude oil and monthly production pattern based on oil imports and exports, national and regional demands, logistics of transportation and other allied matters. In 1981, the work of arriving at demand projection for petroleum products was assigned to the Oil Coordination Committee. The Oil Coordination Committee has recently finalised a report on demand for petroleum products upto 1991-92.

3.4 Distribution policy and planning process and the monitoring of the distribution of petroleum products

(*i*) *Distribution Policy and Planning Process*

(a) The Oil Economy Budget : The Oil Economy Budget is the starting point for the distribution planning process. In accordance with the overall guidance issued by the Department of Petroleum, based on the demand forecasts and in consultation with the Chief Executives of the Oil Companies, the OCC Secretariat draws up for each year an Oil Economy Budget (OEB). The Oil Economy Budget projects production and availability of indigenous crude oil, planned maximum achievable crude throughputs of different refineries, the import requirements of crude oil and types, the total indigenous production of finished refinery products, exports, the overall net requirements of imported finished refinery products and the demand of the various refinery products for the period in question. The basic policy is one of conservation tempered with the need to meet the essential requirements of the industrial and agricultural sectors. Thus, for example, fuel oil releases are strictly governed on the basis of clearances given by the DGTD, to ensure that wherever possible, the utilisation of coal is given preference. HSD which is essential for both the transportation and the agricultural sectors is made available to meet the full demand. Adequate availability of kerosene for lighting and cooking is ensured and the village, town and districtwise distribution is controlled by the State Governments within their allocations. Based on the demands projected in the oil Economy Budget, a sales plan is drawn up apportioning to each of the companies an agreed share of the incremental sales forecast for the year.

(*b*) *The Industry Coordination Meeting*

An industry coordination meeting is held every month to review for the short term covering a span of 4 months, and based on the comprehensive Oil Economy Budget, the production patterns of each refinery are determined, making due allowances for planned/other shutdowns of refineries.

(*c*) *The Supply Plan Meeting*

Under the guidance of the Department of Petroleum, a monthly supply and distribution plan is formulated by the OCC at the end of each month for the following month based on the latest known trends in production, imports, availability and demand. The requirements of different petroleum products for each zone, each state and for the major consumers are determined and a detailed distribution plan drawn up. The supply and distribution plan in regard to movement by the different modes of transportation such as coastally by sea, by rail and by road is also formulated. Statewise allocations of kerosene and HSD are also finalised for each state and for each oil company within the State. The Railway which carries a major portion of the petroleum product traffic are closely associated in the process of drawing up the distribution plan.

(d) Railway Inland Petroleum Movement Committee

At each major refinery loading base, a Railway Inland Petroleum Movement Committee has been formed associating the local railway authorities and members of the oil industry. The agreed distribution plan translated in terms of productwise and destinationwise loading and movement requirements is thereafter implemented by the various oil companies with the assistance and active guidance of Railways Inland Petroleum Movement Committee.

(ii) Monitoring of Distribution of Refinery Products

(a) The Oil Coordination Committee closely monitoring on a continuous basis, the distribution of all refinery products in the country as a whole and in the different states. The OCC Secretriat have evolved a monitoring machinery to ensure effective implementation of thedistribution plan and to see that refinery products are made available in accordance with the plan to meet the demand in the consuming centres.

(b) To assist the OCC Secretariat in monitoring the distribution of refinery products, there has been in existence the organisation of Regional Coordinators, State Level Coordinators, District Level Coordinators, Depot Level Coordinators and Port Coordinators. The Coordinators are officers of designated oil companies functioning on behalf of the entire oil industry in that specific geographic area. The Coordinators keep in close touch with State Government and district level authorities and initiate regular meetings with the local authorities and the oil companies in order to discuss in detail and review such aspects as implementation of the distribution plan, districtwise allocations of State Controlled items such as kerosene, LPG. The State Level Coordinators keep the OCC

Secretariat and the Regional Coordinators advised of all relevant developments. Based on reports received from the State level and the regional Level Coordinators, the OCC interacts, where and if necessary, with the different agencies such as the Railways, the port authorities and the oil companies to evolve alternative contingency plans. The Department of Petroleum is consulted and kept informed and agreed action initiated as called for.

(c) The oil companies themselves at their head offices, regional offices and at other levels operate full fledged supply and distribution departments which get the required feedback relating to the supply and distribution of refinery products within their territory from the depots and divisional offices Both at the regional and the head office level a weekly meeting is held, attended by representatives of all the oil companies to review the supply and distribution situation. Based on these discussions, the Oil Coordination Committee and the Department of Petroleum are advised of the developments and assistance to overcome bottlenecks cleared as and when necessary.

(d) Weekly returns are submitted to the Department of Petroleum, Ministry of Energy, covering significant aspects of sales, stock levels etc. As regards supply position in refinery sensitive areas, even daily reports indicating stocks of products are received by the Department of Petroleum.

4

DISTRIBUTION OF A HEAVY DISTILLATE PRODUCT FROM REFINERIES

Petroleum in the form of bitumen (asphalt) was a common article of commerce at the beginning of civilisation. Asphalt is defined by the American Society for testing materials as "a dark brown to black cementitious material, solid or semisolid in consistency, in which the predominating constituents are bitumen, which occur in nature as such or are obtained as residuals in refining petroleum" (Wellborn, A.S. and J.M. Griffith; 1960). The development of the petroleum industry, beginning in the latter part of the nineteenth century, afforded the means of refining and producing asphaltic materials in huge quantities and in various types and grades under carefully controlled conditions. In the twentieth century as motor transportation developed, the use of bitumen as a road surfacing material developed at an ever increasing rate. Miscellaneous uses of bitumen for industrial and speciality purposes are quite extensive. To mention only a few, asphalts are used in battery cases, automobile undertaking, tyre manufacture, a wide variety of waterproofing and dramp-proofing applications, and as a base for paints and for printing ink (Abraham, H. and V. Nastrand; 1945).

Bitumen has a very important role to play for the development of both the rural and urban sector of the Indian economy. Construction and maintenance of a nation's roads, with the limited funds available, calls for the most economical methods of road construction. Bitumen takes up a substantial part of the cost of surfacing these roads. Economies in road construction necessitate the use of bitumen in bulk. Bulk bitumen essentially means, delivering bitumen in large unit quantities. At present major quantity of bitumen is packed in steel drums at the refineries and despatched to various consuming points. About Rs. 36 crores are being spent annually in steel of which a substantial amount involves foreign exchange. This chapter deals with the determination of the production level of bulk and packed bitumen at various refineries and their distribution to various parts of India in such a manner as to minimize the total cost of production and distribution, taking into account the bulk capacity, packed capacity and production capacity at the refineries and the demand for bulk and packed bitumen at different parts of the country. The period considered is 1984 85, A linear programming model is developed to arrive at the optimum production and distribution pattern for bulk and packed bitumen in India. Also a goal programming model with its ability to incorporate multiple goals is built to lend a finer insight into the problem.

4.1 Importance of Bulk Movement

(i) The Bitumen demand in the country has been far in excess of the availability as a result of which the outstanding order position is building up over the years. Due to limited bulk offtakes and several constraints faced in maintaining packed bitumen supplies, it has become increasingly difficult to utilise even the current total bitumen production capacity. Some of the factors adversely affecting the availability of packed bitumen are :

(a) Limited drumming facilities at the refineries;
(b) Poor availability of wagons for movement of bitumen drums;
(c) Railway booking restrictions;
(d) Irregular/inadequate availability of steel for bitumen drums; and
(e) Labour problems faced at drum plants, filling plants and for drum loading.

(ii) Bulk bitumen is today accepted as an economical and effective method of using this product. Wherever bitumen is produced indigenously, its storage, transportation and application should be in bulk, to the maximum extent. There is a paramount need for maximizing the distribution of bitumen from refineries to the consuming points in bulk form than in packed form in view of :

(a) Bitumen drums for packing are made out of 24G MS sheets, a major quantity of which are imported. It is estimated by the Ministry of Steel and Mines that during 1981-82, the expenditure on these drum sheets amounted to nearly Rs. 36 crores, of which a substantial portion involved foreign exchange.

(b) Leakage during storage/handling/transportation is avoided by using bitumen in bulk. This improves the quality of road work.

(c) Pilferage is more difficult in the case of bulk bitumen.

(d) Packing bitumen in drums to be transported to the work site, then breaking open and discarding such drums increases the ultimate cost of the product and consequently the cost of surfacing the road.

(e) Bulk bitumen is normally delivered hot at a temperature around 150°C perhaps requiring only nominal heating at the work site. Hot delivery of bulk bitumen at site reduces heating costs and helps better utilisation of labour, thus ensuring better progress in daily work. Besides this also reduces consumption of coal and wood, which constitutes a national saving.

(f) Use of bulk bitumen does away with locking up of large funds for packed stocks. Packed bitumen is normally supplied in large quantities at a time. Public works departments may take long time to consume this quantity of bitumen resulting in locking up of funds. In the case of bulk bitumen it can be supplied as and when required in tank lorry loads of specified capacities.

(g) It is estimated that nearly 2% of bitumen packed in a drum gets stuck to the sides which cannot be taken out. This can be avoided by using bulk bitumen. Thus there is substantial scope for conservation of the product.

(h) Timely supplies of bulk bitumen by not having to rely on rail movement help avoid delays. causing project cost escalation.

4.2 Existing System

(i) Currently bulk bitumen is being supplied by refineries at Bombay, Cochin, Madras, Vizag, Haldia, Koyali and Mathura. Table 4.I gives the bulk and packed bitumen upliftment for the years 1978-79 through 1981-82.

(ii) In view of the importance and urgency of conversion of packed bitumen to bulk, a meeting was called by the Secretary (Petroleum) on April 25, 1981 to discuss measures for expeditious promotion of bulk bitumen. Decisions taken at the meeting were that the sale of packed bitumen should be discontinued within 300/400 km of the refinery locations in a period of about two years, and as a long term objective, the use of packed bitumen would be confined only to the hilly areas. Fiscal disincentives to marketing of packed bitumen will also have to be considered. In subsequent meetings held by the Ministry of Petroleum with the Oil Coordination Committee and the industry members, the various states/union territories where bulk can be promoted economically were identified. These states/union territories were asked to prepare plans for accelerated conversion to bulk bitumen use, giving details of their estimated yearwise demand upto 1985-86. A bulk bitumen committee consisting of the oil industry representatives was formed, to recommend steps to be taken to achieve accelerated promotion of bitumen in bulk, in the light of the reports received from states/union territories and report to the oil coordination committee.

4.3 The Approach in this Study

(i) *Problem Parameters*

(a) *Demand estimates*: Demand for packed and bulk bitumen as advised by the states/union territories for the year 1984-85, to the bulk bitumen committee is taken as the basis. Only those states/union territories, where bulk can be promoted economically, as identified by the bulk bitumen committee are considered in the study. Table 4.II summarises the demand of the states/union territories for bulk and packed bitumen for 1984-85.

TABLE 4.I : **Refinerywise Bulk and Packed Bitumen Upliftment Compared to the Total Bitumen Upliftment**

Refinery	1978-79		1979-80		1980-81		1981-82	
	Bulk as % of total	Packed as % of total	Bulk as % of total	Packed as % of total	Bulk as % of total	Packed as % of total	Bulk as % of total	Packed as % of total
BPC Bombay	19	81	18	82	18	82	18	82
HPC Bombay	11	89	11	89	14	86	17	83
CRL (Cochin)	—	—	100	—	100	—	100	—
MRL (Madras)	10	90	14	86	21	79	25	75
Vizag	8	92	9	91	6	94	7	93
Haldia	35	65	51	49	48	52	43	57
Koyali	—	—	12	88	24	76	34	66
Mathura	—	—	—	—	—	—	100	—

Source : Oil Coordination Committee.

TABLE 4.II : **Statewise Bulk/Packed Bitumen Requirements**

(Figures in '000 MT)

State/Union Territory	Bulk	Packed	Total
Andhra Pradesh	26.08	47 2	74.0
Bihar	50.0	105.5	155.5
Chandigarh	2.0	0.4	2.4
Delhi	25.0	2.0	27.0
Gujarat	192.0	24.0	216.0
Haryana	29.2	6.1	35.3
Karnataka	18.3	22.3	40.6
Kerala	32.4	21.6	54.0
Madhya Pradesh	48.3	95.7	144.0
Maharashtra	162.3	40.7	203.0
Orissa	27.5	14.2	41.7
Pondicherry	2.2	0.2	2.4
Punjab	15.0	31.2	146.2
Rajasthan	55.8	26.8	82.6
Tamil Nadu	122.0	33.0	155.0
Uttar Pradesh	94.0	74.0	168.0
West Bengal	141.6	53.8	195.4
	1044.4	598.7	1643.1

Source : Bulk Bitumen Committee's Report (Unpublished).

(*b*) *Supply estimates* : The bulk capacity, packed capacity and the total production capacity at the refineries have been furnished in table 4.III by the oil coordination committee for the year 1984-85.

TABLE 4.III : **Refinerywise Bulk/Packed and Total Production Capacity of Bitumen for the year 1984-85**

(Figures in '000' MT)

Refinery	Bulk Capacity	Packed Capacity	Total Production Capacity
Koyali	140	180	200
Mathura	120	180	300
Haldia	100	120	200
Vizag	15	30	20
Cochin	115	40	150
Madras	140	200	325
Bombay	180	360	600

Source : Oil Coordination Committee.

(*c*) *Price Ex-storage* (*Refinery*) : The total cost of production and distribution considered in this study is arrived at by adding to price ex-storage point (refinery), the costs of distribution. Price ex-storage point (refinery) has been furnisned by the oil coordination committee, after making due allowances for (a) customs/excise duty, (b) marketing margins and profit, (c) drum cost, (d) filling costs, (e) product price adjustment.

TABLE 4.IV : **Price Ex-storage Point (Refinery)**

(*Figures in Rs./MT*)

Refinery	Bulk	Packed
Bombay	1,000	1,400
Vizag	1,000	1,400
Mathura	1,500	1,900
Koyali	1,500	1,900
Cochin	1,500	1,900
Haldia	1,500	1,900
Madras	1,500	1,900

Source : Oil Coordination Committee.

(*d*) *Transportation costs for bulk and packed bitumen* : The modes of transportation considered in this study are (a) Rail for packed, (b) Road for bulk. The cost for packed movement by rail, as furnished by the Oil Coordination Committee is taken as Rs. 70 per MT for a distance of 200 km.

(*e*) By road, bulk bitumen travels in specially built tank lorries. The supply source is the refinery or the bulk bitumen depots. At the refinery hot bitumen is filled into the tank lorries. The quantities are then measured by weight (weighbridge) or volume (dipstick) and the lorries are despatched. A bitumen tank lorry transports bitumen from the refinery (at temperatures of about 150°C) to the worksite and delivers it at stationary Hot-Mix plants—Bulk storage tanks or the bitumen boilers at the work site. Bitumen can then be used, practically with no reheating or nominal reheating as the tank is properly insulated to minimise temperature drop during transit (1 to 2°C per every 2 hours) except in the case of very long haulage during winter months. The tank lorry consists of a truck or trailer chassis with :

(1) Insulated steel tank, usually with heating equipment, manhole on top and discharge arrangements.
(2) Gear rotary pump or similar positive displacement pump with a power take-off from the truck engine or diesel engine for the pump drive.
(3) Distributor or pressure spray bar.
(4) Other appliances like a burner, flexible insulated hose-pipes, thermometer, etc.

The transportation rate is worked out as given in Table-4.V with details furnished by the Oil Coordination Committee.

TABLE 4.V : **Transportation Rate for Bitumen Tank Truck**

A truck can operate 6,000 km. per month i.e. 48,000 km. per year (working season ; 8 months) say 50,000 km per year. Cost of truck @ Rs. 3 lakhs for 10 tonne capacity.

(*a*) *Fixed Cost* :

	Rs.
(1) Interest on investment @ 20%	60,000
(2) Depreciation @ 20%	60,000
(3) Maintenance @ 3%	9,000
(4) Taxes & Insurance @ 4%	12,000
	1,41,000

(*b*) *Operating Cost* :

(1) Fuel : $50{,}000 \text{ km} \times \dfrac{\text{Rs. 3/- per litre}}{\text{4 km per litre}}$	37,500
(2) Servicing and Lubrication	6,000
(3) Cost of Tyre/batteries/accessories	25,000
(4) Salary of one driver/one cleaner inclusive of over-time	25,000
	93,500

Total of (a)+(b)=Rs. 2,34,500

$$\text{Hence Running Cost per MT/km} = \frac{\text{Rs. 2,34,500}}{\text{50,000 km}} \times \frac{1}{\text{10 MT}}$$

$$= \text{Re } 0.47/\text{km/MT}$$

i.e.=Re 0.94/km/MT for return trip.

Taking into account 10% escalation for $1\frac{1}{2}$ years, the transportation cost may be taken as Re 1.10/km/MT.

(*f*) The total cost of production and distribution from each refinery to the centre point of each state is summarised in Table 4.VIII. The same is arrived at by adding the transportation rates to the price ex-storage point (refinery). The distances from the refinery to the centre point of each state/union territory is summarised in Table 4.VI. This distance is taken as the basis for computing the transportation costs for bulk and packed movement (Table 4.VIII). It may be mentioned, that the centre point refers to the graphical centre of each state. The graphical centre is arrived at by taking the central X and Y coordinates for the particular state in question. This is depicted in Figure 4.1. The state capitals are not taken as the basis since the same are not necessarily located at the centre of each state. There are other measures of areal central tendency viz. mean centre, median point and median centre. But since they are time consuming exercises, with only marginal variations in the results, this simple and meaningful technique is adopted in arriving at the centre point of states/union territories.

(*ii*) *The Model*

A linear programming model is developed to arrive at the optimum production and distribution pattern for bulk and packed bitumen in India. The model takes into account the costs of production of bulk and packed bitumen at various refineries, the costs of transportation from the refineries to the consuming points, the bulk, packed and production capacities of bitumen at various refineries and the demand for bulk and packed bitumen at the various consuming points. The modes of transportation considered are road for bulk and by rail for packed. It may be noted from the description of the model that the objective function is to minimise the total cost of production and distribution of bitumen from the refineries to the various consuming points. For each refinery there are three constraints viz., bulk supply capacity, packed supply capacity and the production capacity. Since the anticipated bulk demand is greater than the bulk supply, we have an equal to (=) constraints. Similarly, the anticipated packed demand is less than the packed supply and hence the constraint is of less

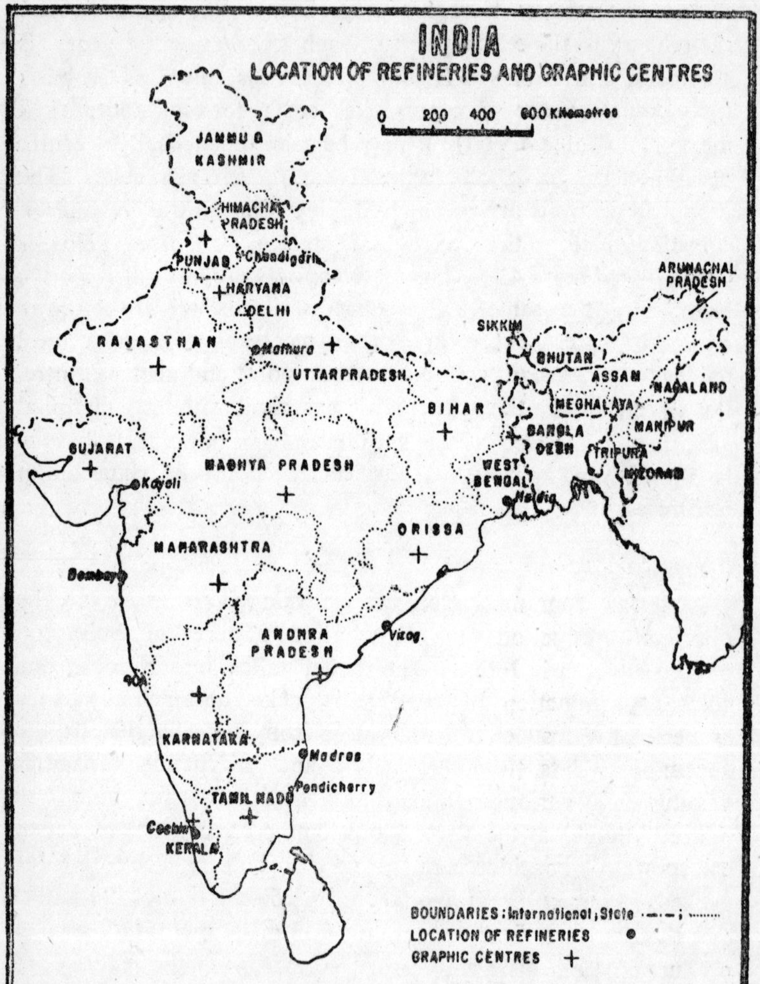
INDIA
LOCATION OF REFINERIES AND GRAPHIC CENTRES
0 200 400 600 Kilometres
JAMMU & KASHMIR
HIMACHAL PRADESH
PUNJAB
Chandigarh
HARYANA
DELHI
RAJASTHAN
Mathura
UTTAR PRADESH
ARUNACHAL PRADESH
SIKKIM
BHUTAN
ASSAM
NAGALAND
MEGHALAYA
MANIPUR
BIHAR
BANGLA DESH
TRIPURA
MIZORAM
GUJARAT
Koyali
MADHYA PRADESH
WEST BENGAL
Haldia
ORISSA
MAHARASHTRA
Bombay
ANDHRA PRADESH
Vizag.
KARNATAKA
Madras
Pondicherry
TAMIL NADU
Cochin
KERALA
BOUNDARIES: International; State
LOCATION OF REFINERIES
GRAPHIC CENTRES

Fig. 4.1

TABLE 4.VI : Distance Matrix

(*Figures in km*)

Refineries	States								
	Andhra Pradesh	Bihar	Chandigarh	Delhi	Gujarat	Haryana	Karnataka	Kerala	Madhya Pradesh
Koyali	1051	1036	989	806	194	796	898	1367	612
Mathura	1286	857	367	143	857	255	1408	1918	592
Haldia	1031	367	1459	1306	1714	1439	1479	1816	929
Vizag	327	816	1561	1357	1347	1469	806	1122	663
Cochin	837	1920	2220	2100	1469	2150	561	71	1398
Madras	357	1418	2000	1750	1398	1825	480	541	1031
Bombay	898	1480	1347	1173	388	1173	592	1031	745

Refineries	States							
	Maharashtra	Orissa	Pondicherry	Punjab	Rajasthan	Tamil Nadu	Uttar Pradesh	West Bengal
Koyali	510	1204	1357	969	449	1408	980	1540
Mathura	897	1061	1725	459	388	1850	327	1112
Haldia	1224	439	1429	1582	1520	1640	949	255
Vizag	714	327	714	1690	1367	939	1122	908
Cochin	1041	1439	429	2310	1850	249	2050	2060
Madras	796	908	115	2030	1650	327	1610	1509
Bombay	408	1224	1082	1357	847	1102	1276	1710

TABLE 4.VII : **Transportation Cost Matrix (Bulk by Road, Packed by Rail)**

(*Figures in Rs/MT*)

Refineries		States								
		Andhra Pradesh	Bihar	Chandigarh	Delhi	Gujarat	Haryana	Karnataka	Kerala	Madhya Pradesh
Koyali	Bulk	1156	1437	1088	887	202	876	988	1503	673
	Packed	368	457	346	282	64	279	314	478	214
Mathura	Bulk	1415	943	404	157	943	281	1549	2110	651
	Packed	450	300	128	50	300	89	493	671	207
Haldia	Bulk	1134	404	1605	1437	1885	1583	1627	1998	1022
	Packed	361	128	511	457	600	504	518	636	325
Vizag	Bulk	360	898	1717	1493	1482	1616	887	1234	729
	Packed	114	285	546	475	471	514	282	392	232
Cochin	Bulk	921	2112	2442	2310	1616	2365	617	78	1538
	Packed	293	672	777	735	514	753	196	25	489
Madras	Bulk	393	1560	2200	1925	1538	2008	528	595	1134
	Packed	125	496	700	613	489	639	168	189	361
Bombay	Bulk	988	1628	1482	1290	427	1290	651	1133	820
	Packed	314	518	471	411	136	411	207	361	261

Table 4.VII—(Contd.)

Refineries		States							
		Maharashtra	Orissa	Pondicherry	Punjab	Rajasthan	Tamil Nadu	Uttar Pradesh	West Bengal
Koyali	Bulk	561	1324	1493	1066	494	1549	1078	1694
	Packed	179	421	475	339	157	493	343	539
Mathura	Bulk	987	1167	1898	504	427	2035	360	1223
	Packed	314	371	603	160	136	648	114	389
Haldia	Bulk	1346	482	1572	1740	1672	1804	1044	280
	Packed	428	154	500	554	532	574	332	89
Vizag	Bulk	785	360	785	1859	1504	1033	1234	999
	Packed	250	114	250	592	478	329	393	318
Cochin	Bulk	1145	1583	472	2541	2035	274	2255	2266
	Packed	364	504	150	809	648	87	718	721
Madras	Bulk	876	999	127	2223	1815	360	1771	1660
	Packed	279	318	40	711	578	114	564	528
Bombay	Bulk	449	1346	1190	1493	932	1212	1414	1881
	Packed	143	428	379	475	296	386	447	599

TABLE 4.VIII : Cost of Production and Distribution Matrix

(Figures in Rs./MT)

Refineries		States								
		Andhra Pradesh	Bihar	Chandigarh	Delhi	Gujarat	Haryana	Karnataka	Kerala	Madhya Pradesh
Koyali	Bulk	1656	2937	2588	2387	1702	2376	2488	3003	2173
	Packed	2268	2357	2246	2182	1964	2179	2214	2378	2114
Mathura	Bulk	2915	2443	1904	1657	2443	1781	3049	3610	2151
	Packed	2350	2200	2028	1950	2200	1989	2393	2571	2107
Haldia	Bulk	2634	1904	3105	2937	3385	3083	3127	3498	2522
	Packed	2261	2028	2411	2357	2500	2404	2418	2536	2225
Vizag	Bulk	1360	1898	2711	2493	2482	2616	1887	2234	1729
	Packed	1514	1685	1946	1875	1871	1914	1682	1792	1632
Cochin	Bulk	2421	3612	3942	3810	3116	3865	2117	1578	3038
	Packed	2193	2572	2677	2635	2414	2653	2096	1925	2389
Madras	Bulk	1393	2560	3200	2925	2538	3008	1528	1595	2134
	Packed	1525	1896	2100	2013	1889	2039	1568	1589	1761
Bombay	Bulk	1988	2628	2482	2290	1427	2290	1651	2133	1820
	Packed	1714	1918	1871	1811	1536	1607	1811	1761	1661

Table 4. VIII—(Contd.)

Refineries		States							
		Maharashtra	Orissa	Pondicherry	Punjab	Rajasthan	Tamil Nadu	Uttar Pradesh	West Bengal
Koyali	Bulk	2061	2824	2993	2566	1994	3049	2578	3194
	Packed	2079	2321	1975	2239	2057	2393	2243	2439
Mathura	Bulk	2487	2667	3398	2004	1927	3535	1860	2723
	Packed	2214	2271	2503	2060	2036	2548	2014	2289
Haldia	Bulk	2846	1982	3072	3240	3172	3304	2544	1780
	Packed	2328	2054	2400	2454	2432	2474	2232	1989
Vizag	Bulk	1785	1360	1785	2395	2504	2033	2234	1999
	Packed	1650	1514	1650	1992	1878	1792	1793	1718
Cochin	Bulk	2645	3083	1972	4041	3535	1774	3755	3766
	Packed	2264	2404	2050	2709	2548	1987	2618	2621
Madras	Bulk	1876	1999	1127	3233	2815	1360	2771	2660
	Packed	1679	1718	1440	2111	1978	1514	1964	1928
Bombay	Bulk	1449	2346	2190	2493	1932	2212	2404	2881
	Packed	1543	1828	1779	1875	1696	1786	1847	1999

than or equal to type ($\leq$). Finally the total of bulk supply capacity and the packed supply capacity should be less than or equal to the total production capacity ($\leq$). Constraints ranging from 1 to 21, refer to the supply constraints of all the seven refineries. Bulk demand cannot be fully met and hence the constraints are of less than or equal to ($\leq$) type. Packed demand can be fully met, and hence the constraints are of equal to ($=$) type. Constraints ranging from 22 to 55 pertain to the demand constraints of the states/union territories.

Objective Function

$$\text{Minimize } Z = \sum_{i=1}^{m} \sum_{j=1}^{n} \alpha_{ij} X_{ij} + \sum_{i=1}^{m} \sum_{j=1}^{n} \beta_{ij} Y_{ij}$$

where

(a) 'α_{ij}' is the total cost per unit of bulk bitumen from the ith refinery to the jth state/union territory.

(b) 'X_{ij}' is the quantity of bulk bitumen to be moved from the ith refinery to the jth state/union territory.

(c) 'β_{ij}' is the total cost per unit of packed bitumen from the ith refinery to the jth state/union territory.

(d) 'Y_{ij}' is the quantity of packed bitumen to be moved from the ith refinery to the jth state/union territory.

Subject to,

(I) Supply constraints of the refineries

Koyali

(1) $\sum_{j=1}^{17} KX_{1,j} = 140$

(2) $\sum_{j=1}^{17} KY_{1,j} \leq 180$

(3) $\sum_{j=1}^{17} KX_{1,j} + \sum_{j=1}^{17} KY_{1,j} \leq 200$

Mathura

(4) $\sum_{j=1}^{17} MTX_{2,j} = 120$

(5) $\sum_{j=1}^{17} MTY_{2,j} \leq 180$

(6) $\sum_{j=1}^{17} MTX_{2,j} + \sum_{j=1}^{17} MTY_{2,j} \leq 300$

Haldia

(7) $\sum_{j=1}^{17} HX_{3,j} = 100$

(8) $\sum_{j=1}^{17} HY_{3,j} \leq 120$

(9) $\sum_{j=1}^{17} HX_{3,j} + \sum_{j=1}^{17} HY_{3,j} \leq 200$

Vizag

(10) $\sum_{j=1}^{17} VX_{4,j} = 15$

(11) $\sum_{j=1}^{17} VY_{4,j} \leq 30$

(12) $\sum_{j=1}^{17} VX_{4,j} + \sum_{j=1}^{17} VY_{4,j} \leq 20$

Cochin

(13) $\sum_{j=1}^{17} CX_{5,j} = 115$

(14) $\sum_{j=1}^{17} CY_{5,j} \leq 40$

(15) $\sum_{j=1}^{17} CX_{5,j} + \sum_{j=1}^{17} CY_{5,j} \leq 150$

Madras

(16) $\sum_{j=1}^{17} MDX_{6,j} = 140$

(17) $\sum_{j=1}^{17} MDY_{6,j} \leq 200$

(18) $\sum_{j=1}^{17} MDX_{6,j} + \sum_{j=1}^{17} MDY_{6,j} \leq 325$

Bombay

(19) $\sum_{j=1}^{17} BX_{7,j}=180$

(20) $\sum_{j=1}^{17} BY_{7,j}\leq 360$

(21) $\sum_{j=1}^{17} BX_{7,j}+\sum_{j=1}^{17} BY_{7,j}\leq 600$

(II) Demand constraints of the States/Union territories

Andhra Pradesh

Bulk

(22) $KX_{1,1}+MTX_{2,1}+HX_{3,1}+VX_{4,1}+CX_{5,1}+MDX_{6,1}$
$+BX_{7,1}\leq 26.8$

Packed

(23) $KY_{1,1}+MTY_{2,1}+HY_{3,1}+VY_{4,1}+CY_{5,1}+MDY_{6,1}$
$+BY_{7,1}=47.2$

Bihar

Bulk

(24) $KX_{1,2}+MTX_{2,2}+HX_{3,2}+VX_{4,2}+CX_{5,2}+MDX_{6,2}$
$+BX_{7,2}\leq 50.0$

Packed

(25) $KY_{1,2}+MTY_{2,2}+HY_{3,2}+VY_{4,2}+CY_{5,2}+MDY_{6,2}$
$+BY_{7,2}=105.5$

Chandigarh

Bulk

(26) $KX_{1,3}+MTX_{2,3}+HX_{3,3}+VX_{4,3}+CX_{5,3}+MDX_{6,3}$
$+BX_{7,3}\leq 2.0$

Packed

(27) $KY_{1,3}+MTY_{2,3}+HY_{3,3}+VY_{4,3}+CY_{5,3}+MDY_{6,3}$
$+BY_{7,3}=0.4$

Delhi

Bulk

(28) $KX_{1,4}+MTX_{2,4}+HX_{3,4}+VX_{4,4}+CX_{5,4}+MDX_{6,4}+BX_{7,4}\leq 25.0$

Packed

(29) $KY_{1,4}+MTY_{2,4}+HY_{3,4}+VY_{4,4}+CY_{5,5}+MDY_{6,4}$ $BY_{7,4}=2.0$

Gujarat

Bulk

(30) $KX_{1,5}+MTX_{2,5}+HX_{3,5}+VX_{4,5}+CX_{5,5}+MDX_{6,5}+BX_{7,5}\leq 192.0$

Packed

(31) $KY_{1,5}+MTY_{2,5}+HY_{3,5}+VY_{4,5}+CY_{5,5}+MDY_{6,5}+BY_{7,5}=24.0$

Haryana

Bulk

(32) $KX_{1,6}+MTX_{2,6}+HX_{3,6}+VX_{4,6}+CX_{5,6}+MDX_{6,6}+BX_{7,6}\leq 29.2$

Packed

(33) $KY_{1,6}+MTY_{2,6}+HY_{3,6}+VY_{4,6}+CY_{5,6}+MDY_{6,6}+BY_{7,6}=6.1$

Karnataka

Bulk

(34) $KX_{1,7}+MTX_{2,7}+HX_{3,7}+VX_{4,7}+CX_{5,7}+MDX_{6,7}+BX_{7,7}\leq 18.3$

Packed

(35) $KY_{1,7}+MTY_{2,7}+HY_{3,7}+VY_{4,7}+CY_{5,7}+MDY_{6,7}+BY_{7,7}=22.3$

Kerala

Bulk

(36) $KX_{1,8}+MTX_{2,8}+HX_{3,8}+VX_{4,8}+CX_{5,8}+MDX_{6,8}+BX_{7,8}\leq 32.4$

Packed

(37) $KY_{1,8}+MTY_{2,8}+HY_{3,8}+VY_{4,8}+CY_{5,8}+MDY_{6,8}$
$+BY_{7,8}=21.6$

Madhya Pradesh

Bulk

(38) $KX_{1,9}+MTX_{2,9}+HX_{3,9}+VX_{4,9}+CX_{5,9}+MDX_{6,9}$
$+BX_{7,9}\leq 48.3$

Packed

(39) $KY_{1,9}+MTY_{2,9}+HY_{3,9}+VY_{4,9}+CY_{5,9}+MDY_{6,9}$
$+BY_{7,9}=95.7$

Maharashtra

Bulk

(40) $KX_{1,10}+MTX_{2,10}+HX_{3,10}+VX_{4,10}+CX_{5,10}+MDX_{6,10}$
$+BX_{7,10}\leq 162.3$

Packed

(41) $KY_{1,10}+MTY_{2,10}+HY_{3,10}+VY_{4,10}+CY_{5,10}+MDY_{6,10}$
$BY_{7,10}=40.7$

Orissa

Bulk

(42) $KX_{1,11}+MTX_{2,11}+HX_{3,11}+VX_{4,11}+CX_{5,11}+MDX_{6,11}$
$+BX_{7,11}\leq 27.5$

Packed

(43) $KY_{1,11}+MTY_{2,11}+HY_{3,11}+VY_{4,11}+CY_{5,11}+MDY_{6,11}$
$+BY_{7,11}=14.2$

Pondicherry

Bulk

(44) $KX_{1,12}+MTX_{2,12}+HX_{3,12}+VX_{4,12}+CX_{5,12}+MDX_{6,12}$
$+BX_{7,12}\leq 2.2$

Packed

(45) $KY_{1,12}+MTY_{2,12}+HY_{3,12}+VY_{4,12}+CY_{5,12}+MDY_{6,12}$
$+BY_{7,12}=0.2$

Punjab

Bulk

(46) $KX_{1,13}+DTX_{2,13}+HX_{3,13}+VX_{4,13}+CX_{5,13}+MDX_{6,13}$
$+BX_{7,13}\leq 15.0$

Packed

(47) $KY_{1,13}+MTY_{2,13}+HY_{3,13}+VY_{4,13}+CY_{5,13}+MDY_{6,13}$
$+BY_{7,13}=31.2$

Rajasthan

Bulk

(48) $KX_{1,14}+MTX_{2,14}+HX_{3,14}+VX_{4,14}+CX_{5,15}+MDX_{6,14}$
$+BX_{7,14}\leq 55.8$

Packed

(49) $KY_{1,14}+MTY_{2,14}+HY_{3,14}+VY_{4,14}+CY_{5,14}+MDY_{6,14}$
$+BY_{7,14}=26.8$

Tamil Nadu

(50) $KX_{1,15}+MTX_{2,15}+HX_{3,15}+VX_{4,15}+CX_{5,15}+MDX_{6,15}$
$+BX_{7,15}\leq 122.0$

Packed

(51) $KY_{1,15}+MTY_{2,15}+HY_{3,15}+VY_{4,15}+CY_{5,15}+MDY_{6,15}$
$+BY_{7,15}=33.0$

Uttar Pradesh

Bulk

(52) $KX_{1,16}+MTX_{2,16}+HX_{3,16}+VX_{4,16}+CX_{5,17}+MDX_{6,16}$
$+BX_{7,16}\leq 94.0$

Packed

(53) $KY_{1,16}+MTY_{2,16}+HY_{3,16}+VY_{4,16}+CY_{5,16}+MDY_{6,16}$
$BY_{7,16}=74.0$

West Bengal

Bulk

(54) $KX_{1,17}+MTX_{2,17}+HX_{3,17}+VX_{4,17}+CX_{5,17}+MDX_{6,17}$
$+BX_{7,17}\leq 141.6$

Packed

(55) $KY_{1,17}+MTY_{2,17}+HY_{3,17}+VY_{4,17}+CY_{5,17}+MDY_{6,17}+BY_{7,17}=53.8$

all $X_{ij}\geqslant 0$ and $Y_{ij}\geqslant 0$

where

K stands for Koyali Refinery
MT stands for Mathura Refinery
H stands for Haldia Refinery
V stands for Vizag Refinery
C stands for Cochin Refinery
MD stands for Madras Refinery
B stands for Bombay Refinery

States/Union territories have been assigned the following numbers

Andhra Pradesh	— 1
Bihar	— 2
Chandigarh	— 3
Delhi	— 4
Gujarat	— 5
Haryana	— 6
Karnataka	— 7
Kerala	— 8
Madhya Pradesh	— 9
Maharashtra	— 10
Orissa	— 11
Pondicherry	— 12
Punjab	— 13
Rajasthan	— 14
Tamil Nadu	— 15
Uttar Pradesh	— 16
West Bengal	— 17

4.4 Suggested Production and Distribution Plan

(i) The production and distribution pattern for 1984-85 as obtained from solution of the model is given in Table 4.IX. Table 4.IX shows the various quantities of bitumen that should be produced at the refineries and the quantities of bulk and packed bitumen that should be distributed from various refineries to the states/union territories.

(ii) Thus it is evident from the results obtained that packed production and movement of bitumen is not desirable from Koyali, Mathura and Cochin refineries, Currently, bulk bitumen is being supplied ex BPC, HPC—Bombay, CRL, MRL, Vizag, Haldia and Koyali refineries. Hence, it is suggested that the proposed commissioning of drumming facilities at the Mathura refinery need not be carried out.

(iii) The results reveal that bulk movement is not possible to Bihar, Chandigarh, Punjab and Rajasthan. For these states/union territories the demand can be met by supplying packed bitumen. Since there is an excess packed supply capacity, any shortfall in bulk supply can be made good by supplying packed bitumen.

(iv) The suggested distribution plan for bulk and packed bitumen, on being compared with the oil coordination committee's tentative plan for 1984-85 reveals significant divergence. The said plan is given in Table 4.X.

(v) The total cost of production and distribution for the distribution plan as per the model is Rs. 2,37,19,52,000. The cost as per the Oil Coordination Committee's tentative distribution plan is Rs. 2,70,52,27,000. Hence the net savings is Rs. 33,32,75,000, if the suggested plan is adopted.

TABLE 4.IX : **Refinerywise and Statewise Production and Distribution Pattern of Bitumen for the year 1984-85**

(*Figure in '000' MT*)

Refinery	Variable	State/Union territory	Bulk	Packed
1. Koyali	$KX_{1,5}$	Gujarat	140	—
Production			140	Nil
2. Mathura	$MTX_{2,4}$	Delhi	25.0	—
	$MTX_{2,6}$	Haryana	29.2	—
	$MTX_{2,16}$	Uttar Pradesh	65.8	
Production			120.0	Nil
3. Haldia	$HX_{3,17}$ & $HY_{3,17}$	West Bengal	100.0	48.7
Production			100.0	48.7

(Contd.)

Table 4.IX (Contd.)

4. Vizag	$VX_{4,11}$	Orissa	15.0	—
	$VY_{4,2}$	Bihar	—	5.0
Production			15.0	5.0
5. Cochin	$CX_{5,8}$	Kerala	32.39	—
	$CX_{5,15}$	Tamil Nadu	82.6	—
Production			114.99	Nil
6. Madras	$MDX_{6,1}$ & $MDY_{6,1}$	Andhra Pradesh	26.8	47.2
	$MDY_{6,2}$	Bihar	—	41.39
	$MDX_{6,7}$ & $MDY_{6,7}$	Karnataka	18.3	22.3
	$MDY_{6,8}$	Kerala	—	21.6
	$MDX_{6,9}$	Madhya Pradesh	6.5	—
	$MDX_{6,10}$	Maharashtra	34.3	—
	$MDX_{6,11}$ & $MDY_{6,11}$	Orissa	12.5	14.2
	$MDX_{6,12}$ & $MDY_{6,12}$	Pondicherry	2.2	0.2
	$MDX_{6,15}$ & $MDY_{6,15}$	Tamil Nadu	39.39	33.0
	$MDY_{6,17}$	West Bengal	—	5.1
Production			139.99	184.99
7. Bombay	$BX_{7,5}$ & $BY_{7,5}$	Gujarat	52.0	24.0
	$BX_{7,10}$ & $BY_{7,10}$	Maharashtra	128.0	40.7
	$BY_{7,2}$	Bihar	—	59.1
	$BY_{7,3}$	Chandigarh	—	0.4
	$BY_{7,4}$	Delhi	—	2.0
	$BY_{7,6}$	Haryana	—	6.1
	$BY_{7,9}$	Madhya Pradesh	—	95.7
	$BY_{7,13}$	Punjab	—	31.2
	$BY_{7,14}$	Rajasthan	—	26.8
	$BY_{7,16}$	Uttar Pradesh	—	74.0
Production			180.0	360.0

TABLE 4.X : **Oil Coordination Committee's Tentative Distribution Plan**

(Figure in '000' MT)

States/Union Territories—rate	Bulk		Packed	
	Refinery	Quantity	Refinery	Quantity
1. Andhra Pradesh	Bombay	26.08	Mathura	47.2
2. Bihar	Vizag Madras	10.00 40.00	Mathura	105.49
3. Chandigarh	—	—	Madras	0.40
4. Delhi	Mathura	24.99	Madras	2.00
5. Gujarat	Koyali	100.00	Koyali Mathura	2.00 22.00
6. Haryana	Mathura	20.00	Madras	6.10
7. Karnataka	Madras	18.3	Koyali Haldia	13.00 11.00
8. Kerala	Cochin	32.4	Koyali	21.59
9. Madhya Pradesh	Bombay	48.3	Bombay Cochin	81.00 10.00
10. Maharashtra	Bombay Koyali	56.00 40.00	Bombay Madras	31.00 10.00
11. Orissa	Vizag	5.00	Haldia	14.20
12. Pondicherry	Madras	2.20	Madras	0.2
13. Punjab	Mathura	15.00	Bombay Madras	20.00 11.00
14. Rajasthan	Bombay	50.00	Bombay	26.80
15. Tamil Nadu	Madras Cochin	82.00 40.00	Madras Bombay	23.00 10.00
16. Uttar Pradesh	Mathura	61.00	Bombay Cochin	49.00 25.00
17. West Bengal	Haldia Cochin	100.00 40.00	Haldia	53.79

4.5 There are numerous problems in the public sector of our economy, which cannot be readily expressed in terms of a single goal. Conflicting multiple goals which are frequently encountered have to be given due consideration. An identification of more elusive and abstract objective functions is of great importance. Goal programming with its ability to incorporate multiple goals has the potential to be applied to problems with financial, social and environmental goals.

(i) The foregoing discussion on the linear programming model considered only one predominant objective criterion, viz. cost minimisation. In the goal programming model the objective function is no longer restricted to a caradinal criterion (i.e. cost); rather it involves criteria like maximization of bulk movement from refineries, meeting the bulk demand as much as possible, and meeting the packed demand. Certainly cost will remain an important decision variable because it determines the resource requirements. However its function will be shifted from that of the objective function to that of a decision constraint.

(ii) In order to achieve the ordinal solution—that is, to achieve the goals according to their importance—negative and/or positive deviations about the goals must be ranked according to the "preemptive" priority factors. In this way the low-order goals are considered only after the high order goals are achieved as desired. If there are goals in several ranks of importance, the preemptive priority factor $P_j(j=1, 2, \ldots, k)$ should be assigned to the negative and/or positive deviational variables. The pre-emptive priority factors have the relationship of $P_1 >>> P_2 >>> P_3 \ldots P_j >> P_{j+1}$, where $>>>$ means "very much greater than". The priority relationship implies that multiplication by n, however large it may be, cannot make the lower-level goal as important as the higher goal (Lee, Sang M., 1972).

(iii) As mentioned earlier, bulk bitumen movement from the refineries is accepted as the most economical and effective method of using this product. In view of the importance of switchover to bulk movement of bitumen, the maximisation of bulk bitumen allocation from the refineries is accorded the first priority (P_1). Second priority is the minimisation of cost of production and distribution (P_2). Third priority is meeting the bulk demand as much as possible (P_3). Last priority is meeting the packed demand of the states/union territories (P_4), in view of the several constraints faced in maintaining packed bitumen supplies.

(iv) One more step to be considered in the goal programming model formulation is the weighting of deviational variables at the same priority level. In the first priority, as

Bombay and Vizag refineries have lower price ex-storage point (refinery), as compared to other refineries, greater weight is attached to movement from Bombay and Vizag refineries. Hence the weights on the basis of price ex-storage point (refinery) for deviational variables 1 and 2 (d_1^- and d_2^-) is 15 and for deviational variables 3 to 7 (d_3^- . . . d_7^- is 10. The deviational variables at the other priority levels have equal weights.

4.6

(*i*) *The Goal Programming Model*

Objective Function

$$\text{Minimize } Z = P_1[15d_1^- + 15d_2^- + 10d_3^- + 10d_4^- + 10d_5^- + 10d_6^- + 10d_7^-] + P_2[d_{49}^-] + P_3[d_8^- + d_9^- + d_{10}^- + d_{11}^- + d_{12}^- + d_{13}^- + d_{14}^- + d_{15}^- + d_{16}^- + d_{17}^- + d_{18}^- + d_{19}^- + d_{20}^- + d_{21}^- + d_{22}^- + d_{23}^- + d_{24}^- + P_4[d_{25}^- + d_{26}^- + d_{27}^- + d_{28}^- + d_{29}^- + d_{30}^- + d_{31}^- + d_{32}^- + d_{33}^- + d_{34}^- + d_{35}^- + d_{36}^- + d_{37}^- + d_{38}^- + d_{39}^- + d_{40}^- + d_{41}^-$$

P_1 : Highest priority is maximisation of bulk allocation from refineries (deviational variables $d_1^-, d_2^-, \ldots, d_7^-$)

P_2 : Second priority is minimisation of cost of production and distribution (deviational variable d_{49}^-)

P_3 : Third priority is meeting the bulk demand as much as possible (deviational variables

$d_8^-, d_9^-, \ldots, d_{24}^-$)

P_4 : Fourth priority is meeting the packed demand. (deviational variables $d_{25}^-, d_{26}^-, \ldots, d_{41}^-$)

Subject to

(I) *Bulk Allocation Constraints fr om Refineries*

Bombay

(1) $X_1+X_2+X_3+X_4+X_5+X_6+X_7+X_8+X_9+X_{10}+X_{11}+X_{12}+X_{13}+X_{14}+X_{15}+X_{16}+X_{17}+d_1^-=180$

Vizag

(2) $X_{35}+X_{36}+X_{37}+X_{38}+X_{39}+X_{40}+X_{41}+X_{42}+X_{43}+X_{44}+X_{45}+X_{46}+X_{47}+X_{48}+X_{49}+X_{50}+X_{51}+d_2^-=15$

Koyali

(3) $X_{69}+X_{70}+X_{71}+X_{72}+X_{73}+X_{74}+X_{75}+X_{76}+X_{77}+X_{78}+X_{79}+X_{80}+X_{81}+X_{82}+X_{83}+X_{84}+X_{85}+d_3^-=140$

Mathura

(4) $X_{103}+X_{104}+X_{105}+X_{106}+X_{107}+X_{108}+X_{109}+X_{110}+X_{111}+X_{112}+X_{113}+X_{114}+X_{115}+X_{116}+X_{117}+X_{118}+X_{119}+d_4^-=120$

Haldia

(5) $X_{137}+X_{138}+X_{139}+X_{140}+X_{141}+X_{142}+X_{143}+X_{144}+X_{145}+X_{146}+X_{147}+X_{148}+X_{149}+X_{150}+X_{151}+X_{152}+X_{153}+d_5^-=100$

Cochin

(6) $X_{171}+X_{172}+X_{173}+X_{174}+X_{175}+X_{176}+X_{177}+X_{178}+X_{179}+X_{180}+X_{181}+X_{182}+X_{183}+X_{184}+X_{185}+X_{186}+X_{187}+d_6^-=115$

Madras

(7) $X_{205}+X_{206}+X_{207}+X_{208}+X_{209}+X_{210}+X_{211}+X_{212}+X_{213}$
$+X_{214}+X_{215}+X_{216}+X_{217}+X_{218}+X_{219}+X_{220}+X_{221}$
$+d_7^-=140$

(II) Bulk Demand Constraints of the States/Union Territories

Andhra Pradesh

(8) $X_1+X_{35}+X_{69}+X_{103}+X_{137}+X_{171}+X_{205}+d_8^-=26.08$

Bihar

(9) $X_2+X_{36}+X_{70}+X_{104}+X_{138}+X_{172}+X_{206}+d_9^-=50.0$

Chandigarh

(10) $X_3+X_{37}+X_{71}+X_{105}+X_{139}+X_{173}+X_{207}+d_{10}^-=2.0$

Delhi

(11) $X_4+X_{38}+X_{72}+X_{106}+X_{140}+X_{174}+X_{208}+d_{11}^-=25.0$

Gujarat

(12) $X_5+X_{39}+X_{73}+X_{107}+X_{141}+X_{175}+X_{209}+d_{12}^-=192.0$

Haryana

(13) $X_6+X_{40}+X_{74}+X_{108}+X_{142}+X_{176}+X_{210}+d_{13}^-=29.2$

Karnataka

(14) $X_7+X_{41}+X_{75}+X_{109}+X_{143}+X_{177}+X_{211}+d_{14}^-=18.3$

Kerala

(15) $X_8+X_{42}+X_{76}+X_{110}+X_{144}+X_{178}+X_{212}+d_{15}^-=32.4$

Madhya Pradesh

(16) $X_9+X_{43}+X_{77}+X_{111}+X_{145}+X_{179}+X_{213}+d_{16}^-=48.3$

Maharashtra

(17) $X_{10}+X_{44}+X_{78}+X_{112}+X_{146}+X_{180}+X_{214}+d_{17}^{-}=162.3$

Orissa

(18) $X_{11}+X_{45}+X_{79}+X_{113}+X_{147}+X_{181}+X_{215}+d_{18}^{-}=27.5$

Pondicherry

(19) $X_{12}+X_{46}+X_{80}+X_{114}+X_{148}+X_{182}+X_{216}+d_{19}^{-}=2.2$

Punjab

(20) $X_{13}+X_{47}+X_{81}+X_{115}+X_{149}+X_{183}+X_{217}+d_{20}^{-}=15.0$

Rajasthan

(21) $X_{14}+X_{48}+X_{82}+X_{116}+X_{150}+X_{184}+X_{218}+d_{21}^{-}=55.8$

Tamil Nadu

(22) $X_{15}+X_{49}+X_{83}+X_{117}+X_{151}+X_{185}+X_{219}+d_{22}^{-}=122.0$

Uttar Pradesh

(23) $X_{16}+X_{50}+X_{84}+X_{118}+X_{152}+X_{186}+X_{220}+d_{23}^{-}=94.0$

West Bengal

(24) $X_{17}+X_{51}+X_{85}+X_{119}+X_{153}+X_{187}+X_{226}+d_{24}^{-}=141.6$

(III) Packed Demand Constraints of the States/Union Territories

Andhra Pradesh

(25) $X_{18}+X_{52}+X_{86}+X_{120}+X_{154}+X_{188}+X_{222}+d_{25}^{-}=47.2$

Bihar

(26) $X_{19}+X_{53}+X_{87}+X_{121}+X_{155}+X_{189}+X_{223}+d_{26}^{-}=105.5$

Chandigarh

(27) $X_{20}+X_{54}+X_{88}+X_{122}+X_{156}+X_{190}+X_{224}+d_{27}^{-}=0.4$

Delhi

(28) $X_{21}+X_{55}+X_{89}+X_{123}+X_{157}+X_{191}+X_{225}+d^-_{28}=2.0$

Gujarat

(29) $X_{22}+X_{56}+X_{90}+X_{124}+X_{158}+X_{192}+X_{226}+d^-_{29}=24.0$

Haryana

(30) $X_{23}+X_{57}+X_{91}+X_{125}+X_{159}+X_{193}+X_{227}+d^-_{30}=6.1$

Karnataka

(31) $X_{24}+X_{58}+X_{92}+X_{126}+X_{160}+X_{194}+X_{228}+d^-_{31}=22.3$

Kerala

(32) $X_{25}+X_{59}+X_{93}+X_{127}+X_{161}+X_{195}+X_{229}+d^-_{32}=21.6$

Madhya Pradesh

(33) $X_{26}+X_{60}+X_{94}+X_{128}+X_{162}+X_{196}+X_{230}+d^-_{33}=95.7$

Maharashtra

(34) $X_{27}+X_{61}+X_{95}+X_{129}+X_{163}+X_{197}+X_{231}+d^-_{34}=40.7$

Orissa

(35) $X_{28}+X_{62}+X_{96}+X_{130}+X_{164}+X_{198}+X_{232}+d^-_{35}=14.2$

Pondicherry

(36) $X_{29}+X_{63}+X_{97}+X_{131}+X_{165}+X_{199}+X_{233}+d^-_{36}=0.2$

Punjab

(37) $X_{30}+X_{64}+X_{98}+X_{132}+X_{166}+X_{200}+X_{234}+d^-_{37}=31.2$

Rajasthan

(38) $X_{31}+X_{65}+X_{99}+X_{133}+X_{167}+X_{201}+X_{235}+d^-_{38}=26.8$

Tamil Nadu

(39) $X_{32}+X_{66}+X_{100}+X_{134}+X_{168}+X_{202}+X_{236}+d_{39}^{-}=33.0$

Uttar Pradesh

(40) $X_{33}+X_{67}+X_{101}+X_{135}+X_{169}+X_{203}+X_{237}+d_{40}^{-}=74.0$

West Bengal

(41) $X_{34}+X_{68}+X_{102}+X_{136}+X_{170}+X_{204}+X_{238}+d_{41}^{-}=53.8$

(*IV*) *Packed Allocation Constraints from Refineries*

Bombay

(42) $X_{18}+X_{19}+X_{20}+X_{21}+X_{22}+X_{23}+X_{24}+X_{25}+X_{26}+X_{27}+X_{28}+X_{29}+X_{30}+X_{31}+X_{32}+X_{33}+X_{34}+d_{42}^{-}=180$

Vizag

(43) $X_{52}+X_{53}+X_{54}+X_{55}+X_{56}+X_{57}+X_{58}+X_{59}+X_{60}+X_{61}+X_{62}+X_{63}+X_{64}+X_{65}+X_{66}+X_{67}+X_{68}+d_{43}^{-}=130$

Koyali

(44) $X_{86}+X_{87}+X_{88}+X_{89}+X_{90}+X_{91}+X_{92}+X_{93}+X_{94}+X_{95}+X_{96}+X_{97}+X_{98}+X_{99}+X_{100}+X_{101}+X_{102}+d_{44}^{-}=180$

Mathura

(45) $X_{120}+X_{121}+X_{122}+X_{123}+X_{124}+X_{125}+X_{126}+X_{127}+X_{128}+X_{129}+X_{130}+X_{131}+X_{132}+X_{133}+X_{134}+X_{135}+X_{136}+d_{45}^{-}=180$

Haldia

(46) $X_{154}+X_{155}+X_{156}+X_{157}+X_{158}+X_{159}+X_{160}+X_{161}+X_{162}+X_{163}+X_{164}+X_{165}+X_{166}+X_{167}+X_{168}+X_{169}+X_{170}+d_{46}^{-}=120$

Cochin

(47) $X_{188}+X_{189}+X_{190}+X_{191}+X_{192}+X_{193}+X_{194}+X_{195}+X_{196}+X_{197}+X_{198}+X_{199}+X_{200}+X_{201}+X_{202}+X_{203}+X_{204}+d_{47}^{-}=40$

Madras

(48) $X_{222}+X_{223}+X_{224}+X_{225}+X_{226}+X_{227}+X_{228}+X_{229}+X_{230}$
$+X_{231}+X_{232}+X_{233}+X_{234}+X_{235}+X_{236}+X_{237}+X_{238}$
$+ d^{-}_{48}=200$

(V) *Cost Constraint*

(49) $1988X_1+2628X_2+2482X_3+2290X_4+1427X_5+2290X_6$
$+1651X_7+2133X_8+1820X_9+1449X_{10}+2346X_{11}$
$+2190X_{12}+2493X_{13}+1932X_{14}+2212X_{15}+2404X_{16}$
$+2881X_{17}+1714X_{18}+1918X_{19}+1871X_{20}+1811X_{21}$
$+1536X_{22}+1811X_{23}+1607X_{24}+1761X_{25}+1661X_{26}$
$+1543X_{27}+1828X_{28}+1779X_{29}+1875X_{30}+1696X_{31}$
$+1786X_{32}+1847X_{33}+1999X_{34}+1360X_{35}+1898X_{36}$
$+2711X_{37}+2493X_{38}+2482X_{39}+2616X_{40}+1887X_{41}$
$+2234X_{42}+1729X_{43}+1785X_{44}+1360X_{45}+1785X_{46}$
$+2859X_{47}+2504X_{48}+2033X_{49}+2234X_{50}+1999X_{51}$
$+1514X_{52}+1685X_{53}+1946X_{54}+1875X_{55}+1871X_{56}$
$+1914X_{57}+1682X_{58}+1792X_{59}+1632X_{60}+1650X_{61}$
$+1514X_{62}+1650X_{63}+1992X_{64}+1878X_{65}+1729X_{66}$
$+1793X_{67}+1718X_{68}+2656X_{69}+2937X_{70}+2588X_{71}$
$+2387X_{72}+1702X_{73}+2376X_{74}+2488X_{75}+3003X_{76}$
$+2173X_{77}+2061X_{78}+2824X_{79}+2993X_{80}+2566X_{81}$
$+1994X_{82}+3049X_{83}+2578X_{84}+3194X_{85}+2268X_{86}$
$+2357X_{87}+2246X_{88}+2182X_{89}+1964X_{90}+2179X_{91}$
$+2214X_{92}+2378X_{93}+2114X_{94}+2079X_{95}+2321X_{96}$
$+1975X_{97}+2239X_{98}+2057X_{99}+2393X_{100}+2243X_{101}$
$+2439X_{102}+2915X_{103}+2443X_{104}+1904X_{105}+1657X_{106}$
$+2443X_{107}+1781X_{108}+3049X_{109}+3610X_{110}+2151X_{111}$
$+2487X_{112}+2667X_{113}+3398X_{114}+2004X_{115}+1927X_{116}$
$+3535X_{117}+1860X_{118}+2723X_{119}+2350X_{120}+2200X_{121}$
$+2028X_{122}+1950X_{123}+2200X_{124}+1989X_{125}+2393X_{126}$
$+2571X_{127}+2107X_{128}+2214X_{129}+2271X_{130}+2503X_{131}$
$+2060X_{132}+2036X_{133}+2548X_{134}+2014X_{135}+2289X_{136}$
$+2634X_{137}+1904X_{138}+3105X_{139}+2937X_{140}+3385X_{141}$
$+3083X_{142}+3127X_{143}+3498X_{144}+2522X_{145}+2846X_{146}$
$+1982X_{147}+3072X_{148}+3240X_{149}+3172X_{150}+3304X_{151}$
$+2544X_{152}+1780X_{153}+2261X_{154}+2028X_{155}+2411X_{156}$
$+2357X_{157}+2500X_{158}+2404X_{159}+2418X_{160}+2536X_{161}$
$+2225X_{162}+2328X_{163}+2054X_{164}+2400X_{165}+2454X_{166}$

$$+2432X_{167}+2474X_{168}+2232X_{169}+1989X_{170}+2421X_{171}$$
$$+3612X_{172}+3942X_{173}+3810X_{174}+3116X_{175}+3865X_{176}$$
$$+2117X_{177}+1578X_{178}+3038X_{179}+2645X_{180}+3083X_{181}$$
$$+1972X_{182}+4041X_{183}+3535X_{184}+1774X_{185}+3755X_{186}$$
$$+3766X_{187}+2193X_{188}+2572X_{189}+2677X_{190}+2635X_{191}$$
$$+2414X_{192}+2672X_{193}+2096X_{194}+1925X_{195}+2389X_{196}$$
$$+2264X_{197}+2404X_{198}+2050X_{199}+2709X_{200}+2548X_{201}$$
$$+1987X_{202}+2618X_{203}+2621X_{204}+1393X_{205}+2560X_{206}$$
$$+3200X_{207}+2925X_{208}+2538X_{209}+3008X_{210}+1528X_{211}$$
$$+1595X_{212}+2134X_{213}+1876X_{214}+1999X_{215}+1127X_{216}$$
$$+3233X_{217}+2815X_{218}+1360X_{219}+2771X_{220}+2660X_{221}$$
$$+1525X_{222}+1896X_{223}+2100X_{224}+2013X_{225}+1889X_{226}$$
$$+2039X_{227}+1568X_{228}+1589X_{229}+1761X_{230}+1679X_{231}$$
$$+1718X_{232}+1440X_{233}+2111X_{234}+1978X_{235}+1514X_{236}$$
$$+1964X_{237}+1928X_{238}+d_{49}^{-}-d_{49}^{+}=1185976.$$

(An arbitrary low cost)

(VI) *Production Capacity Constraints of Refineries*

Bombay

(50) $$X_1+X_2+X_3+X_4+X_5+X_6+X_7+X_8+X_9+X_{10}+X_{11}$$
$$+X_{12}+X_{13}+X_{14}+X_{15}+X_{16}+X_{17}+X_{18}+X_{19}+X_{20}+X_{21}$$
$$+X_{22}+X_{23}+X_{24}+X_{25}+X_{26}+X_{27}+X_{28}+X_{29}+X_{30}+X_{31}$$
$$+X_{32}+X_{33}+X_{34}+d_{50}^{-}=600$$

Vizag

(51) $$X_{35}+X_{36}+X_{37}+X_{38}+X_{39}+X_{40}+X_{41}+X_{42}+X_{43}+X_{44}$$
$$+X_{45}+X_{46}+X_{47}+X_{48}+X_{49}+X_{50}+X_{51}+X_{52}+X_{53}+X_{54}$$
$$+X_{55}+X_{56}+X_{57}+X_{58}+X_{59}+X_{60}+X_{61}+X_{62}+X_{63}+X_{64}$$
$$+X_{65}+X_{66}+X_{67}+X_{68}+d_{51}^{-}=20$$

Koyali

(52) $$X_{69}+X_{70}+X_{71}+X_{72}+X_{73}+X_{74}+X_{75}+X_{76}+X_{77}+X_{78}$$
$$+X_{79}+X_{80}+X_{81}+X_{82}+X_{83}+X_{84}+X_{85}+X_{86}+X_{87}+X_{88}$$
$$+X_{89}+X_{90}+X_{91}+X_{92}+X_{93}+X_{94}+X_{95}+X_{96}+X_{97}+X_{98}$$
$$+X_{99}+X_{100}+X_{101}+X_{102}+d_{52}^{-}=200$$

Mathura

(53) $X_{103}+X_{104}+X_{105}+X_{106}+X_{107}+X_{108}+X_{109}+X_{110}+X_{111}$
$+X_{112}+X_{113}+X_{114}+X_{115}+X_{116}+X_{117}+X_{118}+X_{119}+X_{120}$
$+X_{121}+X_{122}+X_{123}+X_{124}+X_{125}+X_{126}+X_{127}+X_{128}+X_{129}$
$+X_{130}+X_{131}+X_{132}+X_{133}+X_{134}+X_{135}+X_{136}+d_{53}^{-}=300$

Haldia

(54) $X_{137}+X_{138}+X_{139}+X_{140}+X_{141}+X_{142}+X_{143}+X_{144}+X_{145}$
$+X_{146}+X_{147}+X_{148}+X_{149}+X_{150}+X_{151}+X_{152}+X_{153}+X_{154}$
$+X_{155}+X_{156}+X_{157}+X_{158}+X_{159}+X_{160}+X_{161}+X_{162}+X_{163}$
$X_{164}+X_{165}+X_{166}+X_{167}+X_{168}+X_{169}+X_{170}+d_{54}^{-}=200$

Cochin

(55) $X_{171}+X_{172}+X_{173}+X_{174}+X_{175}+X_{176}+X_{177}+X_{178}+X_{179}$
$+X_{180}+X_{181}+X_{182}+X_{183}+X_{184}+X_{185}+X_{186}+X_{187}+X_{188}$
$+X_{189}+X_{190}+X_{191}+X_{192}+X_{193}+X_{194}+X_{195}+X_{196}+X_{197}$
$+X_{198}+X_{199}+X_{200}+X_{201}+X_{202}+X_{203}+X_{204}+d_{55}^{-}=150$

Madras

(56) $X_{205}+X_{206}+X_{207}+X_{208}+X_{209}+X_{210}+X_{211}+X_{212}+X_{213}$
$+X_{214}+X_{215}+X_{216}+X_{217}+X_{218}+X_{219}+X_{220}+X_{221}+X_{222}$
$+X_{223}+X_{224}+X_{225}+X_{226}+X_{227}+X_{228}+X_{229}+X_{230}+X_{231}$
$+X_{232}+X_{233}+X_{234}+X_{235}+X_{236}+X_{237}+X_{238}+d_{56}^{-}=325$

where,

$$X_1, X_2, \ldots, X_{238} \geq 0$$

$$d_1^{-}, d_2^{-}, \ldots, d_{56}^{-} \geq 0$$

$$d_{49}^{+} \geq 0$$

(ii) The computer based solution procedure of goal programming as developed by Lee Sang, M. is adopted (Lee Sang M. 1972). The computer codes for distribution from refinery to the state/union territory is given in Table 4.XI.

TABLE 4.XI : **The Computer Codes for distribution from Refinery to the State/Union Territory**

From / To		Andhra Pradesh	Bihar	Chandigarh	Delhi	Gujarat	Haryana	Karnataka	Kerala	Madhya Pradesh
Bombay	Bulk	X_1	X_2	X_3	X_4	X_5	X_6	X_7	X_8	X_9
	Packed	X_{18}	X_{19}	X_{20}	X_{21}	X_{22}	X_{23}	X_{24}	X_{25}	X_{26}
Vizag	Bulk	X_{35}	X_{36}	X_{37}	X_{38}	X_{39}	X_{40}	X_{41}	X_{42}	X_{43}
	Packed	X_{52}	X_{53}	X_{54}	X_{55}	X_{56}	X_{57}	X_{58}	X_{59}	X_{60}
Koyali	Bulk	X_{69}	X_{70}	X_{71}	X_{72}	X_{73}	X_{74}	X_{75}	X_{76}	X_{77}
	Packed	X_{86}	X_{87}	X_{88}	X_{89}	X_{90}	X_{91}	X_{92}	X_{93}	X_{94}
Mathura	Bulk	X_{103}	X_{104}	X_{105}	X_{106}	X_{107}	X_{108}	X_{120}	X_{110}	X_{111}
	Packed	X_{120}	X_{121}	X_{122}	X_{123}	X_{124}	X_{125}	X_{126}	X_{127}	X_{128}
Haldia	Bulk	X_{137}	X_{138}	X_{139}	X_{140}	X_{141}	X_{142}	X_{143}	X_{144}	X_{145}
	Packed	X_{154}	X_{155}	X_{156}	X_{157}	X_{158}	X_{159}	X_{160}	X_{161}	X_{162}
Cochin	Bulk	X_{171}	X_{172}	X_{173}	X_{174}	X_{175}	X_{176}	X_{177}	X_{178}	X_{179}
	Packed	X_{188}	X_{189}	X_{190}	X_{191}	X_{192}	X_{193}	X_{194}	X_{195}	X_{196}
Madras	Bulk	X_{205}	X_{206}	X_{207}	X_{208}	X_{209}	X_{210}	X_{211}	X_{212}	X_{213}
	Packed	X_{222}	X_{223}	X_{224}	X_{225}	X_{226}	X_{227}	X_{228}	X_{229}	X_{230}

Table 4.XI—(Contd.)

From / To		Maharashtra	Orissa	Pondicherry	Punjab	Rajasthan	Tamil Nadu	Uttar Pradesh	West Bengal
Bombay	Bulk	X_{10}	X_{11}	X_{12}	X_{13}	X_{14}	X_{15}	X_{16}	X_{17}
	Packed	X_{27}	X_{28}	X_{29}	X_{30}	X_{31}	X_{32}	X_{33}	X_{34}
Vizag	Bulk	X_{44}	X_{45}	X_{46}	X_{47}	X_{48}	X_{49}	X_{50}	X_{51}
	Packed	X_{61}	X_{62}	X_{63}	X_{64}	X_{65}	X_{66}	X_{67}	X_{68}
Koyali	Bulk	X_{78}	X_{79}	X_{80}	X_{81}	X_{82}	X_{83}	X_{84}	X_{85}
	Packed	X_{95}	X_{96}	X_{97}	X_{98}	X_{99}	X_{100}	X_{101}	X_{102}
Mathura	Bulk	X_{112}	X_{113}	X_{114}	X_{115}	X_{116}	X_{117}	X_{118}	X_{119}
	Packed	X_{129}	X_{130}	X_{131}	X_{132}	X_{133}	X_{134}	X_{135}	X_{136}
Haldia	Bulk	X_{146}	X_{147}	X_{148}	X_{149}	X_{150}	X_{151}	X_{152}	X_{153}
	Packed	X_{163}	X_{164}	X_{165}	X_{166}	X_{167}	X_{168}	X_{169}	X_{170}
Cochin	Bulk	X_{180}	X_{181}	X_{182}	X_{183}	X_{184}	X_{185}	X_{186}	X_{187}
	Packed	X_{197}	X_{198}	X_{199}	M_{200}	X_{201}	X_{202}	X_{203}	X_{204}
Madras	Bulk	X_{214}	X_{215}	X_{216}	X_{217}	X_{218}	X_{219}	X_{220}	X_{221}
	Packed	X_{231}	X_{232}	X_{233}	X_{234}	X_{235}	X_{236}	X_{237}	X_{238}

(iii) Suggested production and distribution plan

The production and distribution pattern for 1984-85 as obtained from solution of the model is given in Table 4.XII.

TABLE 4.XII : **Suggested production and distribution plan**

(Figure in '000' MT)

Refinery	Computer Code	State/Union Territory	Bulk	Packed
1. Bombay				
	X_5	Gujarat	51.99	
	X_7	Karnataka	18.29	
	X_{10}	Maharashtra	47.82	
	X_1	Andhra Pradesh	26.08	
	X_8	Kerala	6.5	
	X_6	Haryana	29.20	
	X_{32}	Tamil Nadu		33.00
	X_{31}	Rajasthan		26.80
	X_{28}	Orissa		14.20
	X_{29}	Pondicherry		0.20
	X_{33}	Uttar Pradesh		73.99
	X_{30}	Punjab		31.20
	X_{26}	Madhya Pradesh		86.09
PRODUCTION			180	360
2. Vizag				
	X_{45}	Orissa	15.00	
	X_{60}	Madhya Pradesh		5.00
PRODUCTION			15.00	5.00
3. Koyali				
	X_{73}	Gujarat	139.99	
	X_{90}	Gujarat		5.39
	X_{92}	Karnataka		22.29
	X_{91}	Haryana		6.10
	X_{94}	Madhya Pradesh		4.60
	X_{93}	Kerala		21.59
PRODUCTION			132.99	63.00

Table 4.XII—(Contd.)

4. Mathura				
	X_{118}	Uttar Pradesh	93.99	
	X_{106}	Delhi	24.99	
	X_{112}	Maharashtra	0.99	
	X_{121}	Bihar		105.49
	X_{122}	Chandigarh		0.40
	X_{123}	Delhi		2.00
	X_{124}	Gujarat		18.60
	X_{120}	Andhra Pradesh		47.19
PRODUCTION			119.97	173 68
5. Haldia				
	X_{153}	West Bengal	99.99	—
PRODUCTION			99.99	
6. Cochin				
	X_{178}	Kerala	25.80	—
	X_{185}	Tamil Nadu	39.19	
	X_{172}	Bihar	49.99	—
PRODUCTION			114.99	—
7. Madras				
	X_{219}	Tamil Nadu	82.80	—
	X_{216}	Pondicherry	2.20	—
	X_{218}	Rajasthan	54.99	—
PRODUCTION			139.99	—

(iv) There are lot of similarities in the distribution plan between the linear programming and the goal programming models, more so for bulk movement. The total cost of production and distribution as per the goal programming model is Rs 2,69,30,96,000. Though this figure is higher than the one suggested by the linear programming model, it is still less than that of the Oil Coordination Committee's plan. This is understandable as cost minimisation is not the only predominant objective criterion in the goal programming model. Pre-emptive priority factors which are assigned to the deviational variables take cognizance of the fact that high order goals are considered first and must be achieved before low order goals are taken account of. In the model, maximization of bulk allocation at the refineries, is accorded the highest priority and this goal is completely attained. The fourth goal viz. the meeting the packed demand is also fully achieved. There is an over achievement of the second goal i.e. cost. The third goal namely, meeting the bulk demand as much as possible is not attained.

For the petroleum industry the basic decision problem involves the assignment of priorities to various goals and development of programmes to achieve these goals. This is highly desirable because of the critical impact such programmes have on prices, development of backward regions, employment, production and the basic quality of living. Such a decision process constitutes the policy analysis of the industry. Through the application of goal programming the industry will be able to ascertain the soundness of its policies, input requirements for the achievement of set goals, and the degree of goal attainment with the given resources. This review and evaluation process is an integral part of policy analysis. Goal programming provides the optimum solution under the varying inputs and goal structures. The most important advantage of goal programming is its great flexibility, which allows model simulation with numerous variations of constraints and goal priorities. The model allows the decision maker to review critically the priority structure in view of the solution derived by the model. Therefore, goal programming may be particularly well suited for decision analysis in the petroleum industry.

4.7 Investment Analysis

(i) For rapidly increasing the bulk bitumen offtake, and also for achieving the increase in bulk offtake as suggested, additional investment would be required for refinery facilities, transportation and infrastructures at the receiving end.

(ii) Refinery facilities for higher bulk bitumen offtake would require investments for additional filling points, weigh bridges and pumps. An estimated Rs 460 lakhs would be required for providing the necessary facilities at all the refineries handling bulk bitumen (Table 4.XIII).

(iii) Fleets of bitumen tank lorries operated by transport contractors and the consumers themselves, would need to be augmented to meet the increased demand. If new rail fed depots are to be set up, new tank-wagons will have to be fabricated to feed the depots. Also if upcountry depots are planned, bitumen would have to be fed by rail or road to the nearest refinery. From the depots bulk bitumen would be supplied to worksites in tanklorries and operated by transporters or the consumers themselves.

(iv) Receiving facilities at the consumers end primarily consist of mobile storage tanks of different capacities equipped with heating stoves and insulation. The mobile storage tank of either 3 tonne or 6 tonne capacity mounted on a trailer (pneumatic tyres) can be easily hauled from one worksite to the other by a tractor, truck or even a road-roller. If the worksite is in the vicinity of a refinery, this tank can be used for transporting bulk bitumen from the refinery directly to the worksites. Generally when bulk bitumen is used, only mobile site storage tanks should be planned. However, even static bulk bitumen storage tanks of larger capacities could be provided for emergency use, when work stoppage frequently occurs. The storage capacity for this type of tank is dependent on the capacity of storage planned, but the normal capacity of 10/20 MT is both economical and convenient.

(v) Thus it is evident that conversion to bulk movement of bitumen requires enormous investment by the refineries and by the states. Table 4.XIII summarizes the investment plans for the states/union territories for transport and site storage as also the investment for refinery facilities. Table 4.XIV gives the estimated maintenance and repair expenses on the facilities provided. Table 4.XVII details the investment that is required to be made by the bulk bitumen consumers i.e. the PWDs.

(vi) Capital budgeting decisions of the public enterprises, in the final analysis are public industrial investment decisions of the Government and therefore, a number of agencies of the Government are involved in the decision making process. Moreover, the investment decisions for such enterprises are governed by guidelines and directives issued by the Government. Till 1965, no specific instructions or guidelines for capital budgeting decisions by public enterprises were issued by the Government of India. The current capital budgeting decisions for such enterprises are governed by a manual of feasibility studies for public sector projects (Peat, Marwick, Mitchell &

Co. and Planning Commission). The primary factor in deciding whether the project should be approved is the extent to which the proposed project will benefit the nation. The method suggested by the manual for the National Benefit analysis is to reduce to monetary terms as many effects as possible that the project has on the national economy. These effects will be both the benefits derived from the project as well as the costs in national resources that the economy will incur in undertaking the project. This information should then be combined with the relevant commercial data and presented as a comparison of the benefits derived from the project with the costs of the project. An important aspect of the detailed feasibility study of a public sector project suggested by the manual is the analysis of its commercial profitability. Among the methods suggested are the payback period, and the discounted cashflow indices.

The simplest and perhaps the most widely employed quantitative method for appraising capital expenditure decision is the payback period. The payback period refers to the number of years, it will take for the cash benefits to pay the original cost of an investment, normally disregarding salvage value. The payback period can be used as a decision criterion to accept or reject investment proposals. One application of this technique is to compare the actual payback with a predetermined payback, i.e. the payback set up by the management in terms of the maximum period during which the initial investment must be recovered. If the actual payback period is less than the predetermined payback, the project would be accepted, if not, it would be rejected. The distinguishing characteristic of the discounted cash flow capital budgeting technique is that they take into consideration the time value of money while evaluating the costs and benefits of a project. The present value method is one of the discounted cash flow (DCF) or time-adjusted techniques. The method requires the cash flow to be discounted at a certain rate, known as the cost of capital. Net present value may be defined as the summation of the cash proceeds in each year minus the summation of present values of the net cash outflows in each year. The decision rule for a project under net present value is to accept the project if net present value is positive and reject if it is negative. The internal rate of return is the discount rate which equates the aggregate present value of net cash inflows with the aggregate present value of cash outflows of a project.

(vii) Net present value (NPV) from the national point of view has been calculated in Table 4.XVI. The net present value is positive, based on the desired rate of return established by the Reserve Bank of India and the Planning Commission, the return being 12%. This figure has been used as the discount rates in the present value calculations. Also the internal rate of return (IRR) as per Table 4.XVII is 24.8%. Thus the IRR is twice the desired rate of return. Hence from the national point of view, investments for switchover to bulk movement of bitumen is justified. Table 4.XV summarises the savings per MT of bulk bitumen over packed bitumen. The payback period for a PWD division is 1.24 years. Also the net present value from the consumers point of view is positive as per Table 4.XX.

4.8 Conclusion

(i) Inspite of the substantial financial benefits in use of bulk bitumen, many government consumers located nearby to the refineries continue to use large quantities of packed bitumen. To ensure the conversion to bulk bitumen, mandatory bulk zones should be introduced i.e. divisions located within a specified distance will have to be supplied only bulk bitumen. Packed supply to divisions in the mandatory bulk zones should be limited to a nominal quantity for patch repair work etc.

(ii) Packed bitumen supplies should be made only to hilly and other inaccessible areas.

(iii) The facilities required at the refineries to meet the higher demand of bulk bitumen and to ensure speedy turn-around of tanklorries should be implemented urgently. A bitumen tanklorry takes a lot of time due to factors other than the actual filling operation at some of the refineries. Even though the actual filling time is fifteen minutes for a 10 MT tanklorry, at some of the refineries, the tanklorry takes almost 2 to 4 hours from the time it gets into the refinery till the time leaves for its destination after completing all formalities. Thus the tanklorry utilisation is adversely affected. It is strongly recommended that additional facilities, viz., filling loses, weigh bridge and manpower should be designed in a scientific manner, so that a bitumen tanklorry takes no longer than one hour from the time of reporting to the time of departure for destination at any refinery. This would go a longway in increasing the off-takes by optimum utilisation of the available tanklorry fleet.

(iv) The state governments should set up suitable monitoring systems to review periodically the progress in reaching their bulk bitumen allocations from year to year. Ministry of Shipping and Transport (Roads Wing) should also set up a continuous system of monitoring the progress made by different states and assist them in taking corrective action. Wherever possible Disincentives to defaulting states may also be considered.

(v) In view of the attractive economic return to the consumers of bulk bitumen, urgent action must be initiated by the states to instal the facilities for the use of bulk bitumen. The oil industry development board has offered monetary assistance to the states. Meanwhile, the Planning Commission is also likely to sanction additional funds to the states for purchase of bulk storage and handling equipment. Ministry of Petroleum has already agreed to arrange priority allotment of tank-truck chassis. The National highways construction/maintenance estimates are presently being worked out by the states on the basis of packed bitumen cost. Ministry of Shipping and Transport, may consider specifying the use of bulk bitumen while sanctioning these estimates. Since the actual cost of the project will then be lower, the money so saved may be earmarked for installation of bulk bitumen facilities.

(vi) The installation of rail-fed depots could be considered wherever necessary. However, setting up of a new rail fed depot would take at least two years. In this case, projects will have to be developed in consultation with the Railways. Setting up of road-fed depots will be much quicker and more economical than setting up of rail-fed depots (Table 4.XXI & 4.XXII). Experience gained in Tamil Nadu shows that road fed bulk bitumen depots, constructed well within the economic road-fed zone of the refinery (nearly 400 km) have not proved to be much useful, since generally all the supplies around these depots are made direct to the work site from Madras and Cochin refineries. Hence, these depots, which are almost redundant, are presently being used only as buffer or storage to be used in the event of tanklorry breakdown or refinery shut-downs. Therefore, road-fed depots should be located over 400/500 Km from the supply point on a selective basis. Since investment on these depots are Rs. 20 lakhs per depot, state governments may like to put up depots at a few places.

(vii) Adequate manpower should be provided both at the refinery points and in the marketing offices so that the necessary service can be provided to the bitumen consumers for conversion to bulk. The state PWD may also consider provision of necessary man power to pursue the conversion to bulk bitumen and monitoring of projects/bulk bitumen offtakes.

(viii) Oil industry should arrange to publish comprehensive literature and prepare new films on bulk bitumen to facilitate the education of the personnel involved in the use of bulk bitumen. Necessary seminars/group discussions at frequent intervals could be organised for each PWD/NH Circle.

TABLE 4.XIII : **Statewise Yearwise Investment Plans**

(*Figure in Rs. Lakhs*)

	1982-83	1983-84	1984-85	1985-86	Total
(*A*) *Transport & Site Storage*					
Andhra Pradesh	26	110	120	65	321
Bihar	62	98	61	141	362
Chandigarh	7	—	—	—	7
Delhi	4	—	—	—	4
Gujarat	384	98	154	221	857
Haryana	166	77	39	—	282
Karnataka	—	55	65	—	120
Kerala	104	—	—	—	104
Madhya Pradesh	50	75	93	—	218
Maharashtra	147	23	—	—	170
Orissa	106	6	4	—	116
Pondicherry	15	4	—	—	19
Punjab	12	43	51	115	221
Rajasthan	53	52	56	51	212
Tamil Nadu	27	165	180	130	502
Uttar Pradesh	336	190	112	247	885
West Bengal	167	19	19	1	206
Total	1666	1015	954	971	4606

(Contd.)

(*B*) *Refinery Facilities*	—	220	240	—	460
Total (A)+(B)	1666	1235	1194	971	5066

Note : (1) Investment on tank lorries under the head Transport and Site Storage is based on the estimated availability of contractors fleet.

(2) Investment on site facilities is dependent on the number of PWD Divisions and the number of sub-divisions/work sites per division for each state.

Source : Oil Coordination Committee.

TABLE 4.XIV : Yearwise Maintenance and Repair Expenses on Facilities Provided

(*Figure in Rs. Lakhs*)

	Investment	1982-83	1983-84	1984-85	1985-86	Total
1982-83	1666	50	—	—	—	50
1983-84	1235	50	37	—	—	87
1984-85	1194	50	37	36	—	123
1985-86	971	84	37	36	29	186
1986-87		84	62	36	29	211
1987-88		84	62	60	29	235
1988-89		167	62	60	49	338
1989-90		167	124	60	49	400
1990-91		167	124	120	49	460
1991-92		167	124	120	97	508

Note : (i) Maintenance and Repair expenses per year assumed
@ 3% for the first three years
@ 5% for the next three years
@ 10% for the balance period.

Source : Oil Coordination Committee.

TABLE 4.XV : Savings on National Basis

(*Per MT of Bulk Bitumen*)

1.	Drum Steel Cost	Rs. 360.00
2.	Filling Cost	Rs. 20.00
3.	Handling Charges at Rail Head	Rs. 20.00
4.	Transportation Charges from Stores to Worksite	Rs. 50.00
5.	Heating at Worksite	Rs. 20.00
6.	Storage/Handling Facilities at Worksite	Rs. (29)
7.	Bitumen Product Loss	Rs. 30.00
8.	Freight Differential (say for 200 Km distance rail/road)	Rs. (150)
	Net Savings per MT of Bulk Bitumen	Rs. 321.00

Source : Oil Coordination Committee.

TABLE 4.XVI : **Discounted Cash Flow : National Point of View**

(Net Present Value Computation (NPV)

Year	Total Estimated Bulk Sales (000 MT)	Incremental Bulk Sales Due to Additional Facilities (Col. 2—350) (000 MT)	Cash In-Flow (Col. 3 × 321) (Rs. Lakhs)	Cash Outflow for Facilities (Rs. Lakhs)	Cash Outflow for M & R (Rs. Lakhs)
(1)	(2)	(3)	(4)	(5)	(6)
1982-83	400	50	161	1666	50
1983-84	500	150	482	1235	87
1984-85	625	275	883	1194	123
1985-86	700	340	1124	971	186
1986-87	800	450	1445	—	211
1987-88	900	550	1766	—	235
1988-89	975	625	2006	—	338
1989-90	1050	700	2247	—	400
1990-91	1150	800	2568	—	460
1991-92	1200	850	2729	—	508

Table 4.XVI (*Cont.*)

Year	Total Cash Outflow	Discount Rate 12% Discount Factor	Present Value of Cash Inflow Discount Rate 12%	Present Value of Cash Outflow Discount Rate 12%
	(7)	(8)	(9)	(10)
1982-83	1716	.893	143.77	1532.38
1983-84	1322	.797	384.15	1053.63
1984-85	1317	.712	628.69	937.70
1985-86	1157	.636	714.86	735.85
1986-87	211	.567	819.31	119.63
1987-88	235	.507	895.36	119.14
1988-89	338	.452	906.71	152.77
1989-90	400	.404	907.78	161.6
1990-91	460	.361	927.04	166.06
1991-92	508	.322	878.73	163.57
			7206.4	5142.33

Net present value (NPV)=Present value of cash inflows−Present value of cash outflows
Hence NPV =2064.07
NPV>0

*Planning Commission assumes cost of capital as 12% for appraisal of investments.

TABLE 4.XVII : **Discounted Cash Flow : National Point of View**

(*Internal Rate of Return* (*IRR*) *Computation*)

Year	Total Estimated Bulk Sales (000 MT)	Incremental Bulk Sales Due to Additional Facilities (Col. 2−350)	Cash Inflow (Col. 3×321) (Rs. Lakhs)	Cash Outflow for Facilities (Rs. Lakhs)	Cash Outflow for M & R (Rs. Lakhs)
(1)	(2)	(3)	(4)	(5)	(6)
1982-83	400	50	161	1666	50
1983-84	500	150	482	1235	87
1984-85	625	275	883	1194	123
1985-86	700	350	1124	971	186
1986-87	800	450	1445	—	211
1987-88	900	550	1766	—	235
1988-89	975	625	2066	—	338
1989-90	1050	700	2274	—	400
1990-91	1150	800	2568	—	460
1991-92	1200	850	2729	—	508

Table 4.XVII (*Contd.*)

Year	Total Cash Outflow	Net Cash Inflow	24% Factor	Rs. Lakhs	25% Factor	Rs. Lakhs
	(7)	(8)	(9)	(10)	(11)	(12)
1982-83	1716	(1555)	.806	(1253)	.800	(1244)
1983-84	1322	(840)	.650	(546)	.640	(538)
1984-85	1317	(434)	.524	(227)	.512	(222)
1985-86	1157	(33)	.423	(14)	.410	(14)
1986-87	211	1234	.341	421	.328	405
1987-88	235	1531	.275	421	.262	401
1988-89	338	1728	.222	384	.210	363
1989-90	400	1847	.179	331	.168	310
1990-91	460	2108	.144	304	,134	282
1991-92	508	2221	.116	258	.107	238
				+ 79		− 19

Internal Rate of Return (IRR)=24.8%

TABLE 4.XVIII : **Facilities/Investment Required for Using Bulk Bitumen**

1. Definition

Customer : PWD (Public Works Department) Division

Offtake : 1500 MT per Annum (Based on average upliftment)

Location : 200 Km from Refinery

2. Facilities

A. *Transport*

Monthly Bitumen Requirement	=188
Lorry Capacity (MT)	= 12
No. of trips per month a lorry can make	= 16
No. of tank lorry trips required	= 16
No. of tank lorries required	= 1

B. *Work Site Storage Facility*

Storage capacity requirement per PWD division	= 36 MT
No. of 3 MT capacity mobile storage tanks	= 8
No. of 6 MT capacity mobile storage tanks	= 2

3. Investment

A. Transport	: 1 No. : 12 MT Capacity Tank lorry @ Rs. 3,00,000	=Rs. 3,00,000
B. Site Storage	: 8 No. : 3 MT Capacity Mobile storage tank @ Rs. 25,000	=Rs. 2,00,000
	–2 No. : 6 MT Capacity Mobile storage tank @ Rs. 35,000	=Rs. 70,000
		=Rs. 2,70,000
Total Investment (A)+(B)		=Rs. 5,70,000

Source : Oil Coordination Committee

TABLE 4.XIX : **Computation of Payback Period for a PWD Division**

A. Investment		=Rs. 5,70,000 i.e. Rs. 380 per MT of Bulk Bitumen
B. Savings per MT of Bulk Bitumen		
	Savings (Rs)	
(1) Price Ex-refinery	500	
(2) Sales Tax @ 5%	25	
(3) Filling Cost	20	
(4) Handling charges at rail road	20	
(5) Transportation charges from stores to worksite	50	
(6) Heating at worksite	20	
(7) Storage handling facilities at worksite	(29)	
(8) Bitumen product loss (savings on bitumen stuck to drums)	30	
(9) Resale value of drums	(180)	
(10) Freight differential (for 200 Km distance rail/road)	(150)	
Net Savings		306 per MT of Bulk Bitumen
Payback period	$=\frac{\text{Rs. } 380}{\text{Rs. } 306}=1.241$	

TABLE 4.XX : **Discounted Cash Flow : Consumers Point of View**

(*Net Present Value (NPV) Computation*)

Year	Total Estimated Bulk Sales (000 MT)	Incremental Bulk Sales Due to Additional Facilities (000 MT)	Cash Inflow (Col. 3 × 306) (Rs. Lakhs)	Cash Outflow for Facilities (Rs. Lakhs)	Cash Outflow for M & R (Rs. Lakhs)
(1)	(2)	(3)	(4)	(5)	(6)
1982-83	400	50	153	1666	50
1983-84	500	150	459	1235	87
1984-85	625	275	842	1194	123
1985-86	700	350	1071	971	186
1986-87	800	450	1377	—	211
1987-88	900	550	1683	—	235
1988-89	975	625	1913	—	338
1989-90	1050	700	2142	—	400
1990-91	1150	800	2448	—	460
1991-92	1200	850	2601	—	508

Table 4.XX (*Contd.*)

Year	Total Cash Outflow	Discount Rate 12% Discount Factor	Present Value of Cash Inflows Discount Rate 12%	Present Value of Cash Outflows Discount Rate 12%
	(7)	(8)	(9)	(10)
1982-83	1716	.893	136.62	1532.38
1983-84	1322	.797	365.82	1053.63
1984-85	1317	.712	599.50	937.70
1985-86	1157	.636	681.15	735.85
1986-87	211	.567	780.75	119.63
1987-88	235	.507	853.28	119.14
1988-89	338	.452	864.67	152.77
1989-90	400	.404	865.36	161.6
1990-91	460	.361	883.72	166.06
1991-92	508	.322	837.52	163.57
			6868.39	5142.33

Net Present Value (NPV)=1726.06
NPV >0

TABLE 4.XXI : **Cost of a Rail Fed Depot Served by Rakes Ex-Refinery**

Item	*Amount* Rs./Lakhs).
1. Civil works	30.0
2. Tankage 2500 MTs	13.0
3. Product pipeline works	10.0
4. Steam pipe work	3.0
5. Yard lighting & electrification	3.0
6. Cost of material	71.0
Total :	130.0

Source : Oil Coordination Committee, New Delhi.

TABLE 4.XXII : **Cost of Road-Fed Depot**

Items	*Amount* (Rs./lakhs)
1. Civil works	4.0
2. Tankage 500 M. Tons	5.0
3. Product pipeline work	1.5
4. Yard lighting and electrification	2.0
5. Cost of material	7.5
Total :	20.0

Source : Oil Coordination Committee, New Delhi.

5

COASTAL MOVEMENT OF LIGHT, MIDDLE AND HEAVY DISTILLATE PRODUCTS

The preceding chapter gave an insight into the distribution aspects of a heavy distillate product from the refineries. The present section deals with the coastal movement of refinery products. For this study light, middle and heavy distillate products are considered. The purpose of the study is to establish the distribution pattern of the ports in the country for light middle and heavy distillate products and thereby deduce the port traffic. Based on the distribution pattern, an attempt is to be made to identify the problem areas with regard to the facilities required for handling the anticipated traffic at various port locations. Since light, middle and heavy distillate products are considered, recourse is taken to a multiproduct distribution model.

5.1 The Role of Coastal Movement in the Petroleum Industry

The real significance of ocean-borne petroleum movement has been known to oilmen for almost as many years as the industry itself. Taking the foreign area in total and excluding the Russian bloc countries, almost eight barrels out of every ten barrels of oil consumption move along an ocean trade route between oil well and the final consumer (Kohle, L.F. and A.J. Kelly; 1959). In Europe the figure is still higher—over 90% dependency on ocean transport. The advantages of transporting petroleum by water as against railway transport stem from the greater capacity of petroleum tankers, the simplicity of their loading and unloading and its cheapness. For this reason, petroleum was and is transported by water in all cases when it is produced near sea Costs or navigable waters or when petroleum processing plants are located near water routes. A country's petroleum resources are unevenly distributed; as a rule, the centres of the work intensive production of petroleum and natural gas do not coincide with the centres of highest consumption. This disproportionate location of petroleum producing and consuming centres led to the necessity of transporting petroleum by sea (Sokolov, V.; 1972).

Coastal shipping is generally accepted as the most energy efficient and cheapest mode of transport for carriage of bulk traffic over longer distances. (Report of the National Transport Policy Committee, Planning Commission, Govt. of India 1980).

Thus as we look back over the role that marine transportation has played throughout development of the oil industry, we find that the continuous search for adequate and economical means of transporting the ever increasing quantity of petroleum products from the sources to the points of consumption has led to the development of modern, efficient tanker. (Kahle and Kelly; 1959).

From the first simple design, the tanker has grown until it has become one of the most highly developed and efficient contemporary ship types constituting as it does, more than one-fifth of the worlds merchant ship tonnage. The tanker's part in the business of petroleum is a vital one. Tankers are conventionally classified as "Clean" or "dirty" according to whether they are carrying "white" oils such as motor spirit or kerosene or "black" oils such as crude or fuel oil.

5 2

(i) *Existing System* : In India a total number of eleven ports handle 40 million tonnes of crude and finished products. Keeping in line with the national development and the energy need, ambitious plans have been drawn up to increase the refining capacities of existing refineries which will result in the increased traffic of petroleum cargoes at the ports. The increasing trend of petroleum products traffic at the ports is revealed in Table 5.I.

(ii) A brief description about the ports will be pertinent in the present context.

(a) *Bombay* : Port of Bombay is the largest oil port of the country. Bombay being the landfall for ONGC submarine

TABLE 5.I : **Share of Traffic of Petroleum Products in the Total Traffic at Major Ports During 1965-66 to 1980-81**

(*Figure in Million Tonnes*)

Name of the Port	Year	Total Traffic	POL Traffic	POL as % age of Total Traffic at the Port
Calcutta/ Haldia	1965-66	9.84	1.53	15.6
	1969-70	6.89	1.55	22.5
	1973-74	6.33	1.97	31.1
	1977-78	7.55	3.41	42.2
	1978-79	7.98	3.92	49.1
	1979-80	8.56	4.58	53.5
	1980-81	9.29	5.15	55.4
Bombay	1965-66	18.12	9.35	51.6
	1969-70	15.04	8.60	57.2
	1973-74	18.45	11.05	59.9
	1977-78	16.90	9.86	58.3
	1978-79	15.60	7.83	50.2
	1979-80	16.57	7.84	47.3
	1980-81	16.73	8.18	48.9
Madras	1965-66	4.87	0.93	19.1
	1969-70	6.44	2.24	34.8
	1973-74	7.75	3.47	44.8
	1977-78	8.07	3.47	43.0
	1978-79	9.57	3.63	37.9
	1979-80	9.99	3.73	37.3
	1980-81	10.40	4.58	44.0
Cochin	1965-66	2.87	1.04	36.2
	1969-70	4.79	3.56	74.3
	1973-74	3.72	2.50	67.2
	1977-78	5.17	3.81	37.7
	1978-79	5.47	3.84	70.2
	1979-80	5.47	3.87	70.8
	1980-81	5.27	3.94	74.8
Kandla (including Vadinar)	1965-66	2.51	0 93	37.1
	1969-70	2.11	0.98	46.4
	1973-74	3.12	1.50	51.3
	1977-78	3.82	2.75	72.0

	1978-79	5.87	4.37	74.4
	1979-80	7.27	5.45	75.0
	1980-81	8.82	6.82	77.3
Visakh	1965-66	4.38	2.01	45.9
	1969-70	8.30	1.94	23.4
	1973-74	7.95	1.93	24.3
	1977-78	9.69	1.74	18.0
	1978-79	10.05	1.94	19.3
	1979-80	10.23	1.94	19.0
	1980-81	10.26	2.39	23.3
Tuticorin	1977-78	0.63	0.35	55.6
	1978-79	1.06	0.32	30.2
	1979-80	2.41	0.44	18.3
	1980-81	2.63	0.45	17.1
Mormugao	1965-66	7.86	0.09	1.1
	1973-74	14.34	0.55	4.5
	1977-78	11.28	0.62	5.5
	1978-79	10.80	0.64	5.9
	1979-80	14.80	0.73	4.9
	1980-81	13.76	0.59	4.3
Mangalore	1977-78	0.38	0.20	52.6
	1978-79	0.87	0.19	21.8
	1979-80	0.99	0.21	23.3
	1980-81	0.96	0.22	22.9
Grand Total	1965-66	50.46	13.88	31.5
	1969-70	54.35	19.02	35.0
	1973-74	63.96	23.07	36.1
	1977-78	66.23	26.21	39 6
	1978-79	69.44	26.68	38.4
	1979-80	78.49	28.78	36.7
	1980-81	80.41	32.30	40.2

Source : Indian Petroleum and Petrochemical Statistics—1981-82
Ministry of Petroleum, Chemicals & Fertilizers, Govt, of India

pipeline, the entire Bombay High production is first brought to Bombay and then after meeting the refinery requirements, the balance of Bombay High production is shipped out of Bombay to other Coastal refineries. Development of this port has special significance and impact on the rest of the oil ports of the country.

(b) *Kandla* : The port of Kandla acts as a bridging point for movement of petroleum products to the deficient North-west sectors of the country. With Mathura refinery going in full stream and proposed construction of Karnal refinery, the surplus products are envisaged to move out through Kandla. Limits of Kandla Port Trust were extended to Vadinar, thus enabling Vadinar to be a major port of Kandla. Vadinar is situated 60 nautical miles from the Kandla Harbour.

(c) *Okha* : Okha being a minor port comes under the jurisdiction of Gujarat Maritime Board. Presently the port handles limited quantity of petroleum products to meet the local requirements of the area.

(d) *Goa* : Port of Mormugao has a limited oil traffic. But considering the available draft, this port is used as a lighterage point for imported products on MR size vessels which hitherto had been sent to Bombay.

(e) *Mangalore* : Mangalore has been selected as a site for location of a new refinery of 6 m.t.p.a. capacity. The existing draft and the berth of the port is adequate to meet the present requirement of petroleum products of the port.

(f) *Cochin* : Cochin port has two oil berths namely North and South Jetties. Expansion of the port facilities are in progress to handle the increasing petroleum products traffic at this port.

(g) *Tuticorin* : Tuticorin is primarily used by tankers carrying Nephtha and Furnace oil for the local requirement of the area. Ships loaded upto 30′ can be handled at this port.

(h) *Madras* : There is only one oil jetty at Madras in the Bharati Docks. Presently Madras Port handles 5 m.t.p.a. of crude and petroleum products.

(i) *Vizag* : Presently 2.2 m.t.p.a. of petroleum products are being handled through Vizag Port. With the expansion of Vizag refinery by 1984, crude traffic will take an upward swing to the tune of 4.5 m.t.p.a.

(j) *Haldia :* There is only one jetty at Haldia to handle crude and petroleum products. The drafts at Haldia though designed earlier for 40′ have not gone beyond 25′ to 30′. Government is now actively considering construction of additional oil jetty at Haldia.

(iii) In ports of **Bombay, Cochin,** Madras, Vizag and Haldia, petroleum products can be loaded as well as unloaded. Kandla, Okha, Mangalore, Goa and Tuticorin ports can only receive the products, i.e. products can be unloaded only.

Products are moved from port to port by tankers of varying capacity. Table 5.II summarizes the Indian tanker fleet as on 1. 4. 1983.

TABLE 5.II : **Indain Tanker Fleet as on 1.4.1983**

Owner	Name of the Ship	DWT Tonnes (in Tonnes)	Year of Build
1. Shipping Corporation. of India (35)	1. Maharashi Dayanand*	123450	1978
	2. Vishvesvarya	16450	1973
	3. Sarojini Naidu	16440	1975
	4. Bhagat Singh	16540	1974
	5. Rafi Ahmed Kidwai	24439	1975
	6. Jain Narayan Vyas	24529	1975
	7. Aurobindo	24411	1976
	8. Kolandia	24490	1976
	9. Desh Bandhu	36790	1974
	10, Lajpat Rai	49337	1965
	11. Gargi	58739	1963
	12. Jawaharlal Nehru	87170	1969
	13. Lal Bahadur Shastri	87170	1970
	14. Satyamurti	89351	1975
	15. Vivekananda	89371	1974
	16. Netaji Subhash Bose	89392	1973
	17. Lokmanya Tilak	89411	1975
	18. B.R. Ambedkar	89454	1974
	19. Chhatrapati Shivaji	89490	1974
	20. Rajendra Prasad	115723	1975
	21. Zakir Hussain	115853	1975
	22. Kanchanjunga	276755	1975
	23. Koyali	276800	1976
	24. Dadabhai Naoroji	24400	1977
	25. Bhaskara*	21941	1965

Table 5.II (*Contd.*)

	26. Chanakya*	21962	1965
	27. Leelavati*	21962	1965
	28. Dailadila*	86890	1971
	29. Bellary*	86959	1970
	30. Barauni*	87200	1970
	31. Vallabhabhai Patel*	113920	1977
	32. Moti Lal Nehru	114086	1973
	33. Vallathol	114152	1976
	34. Abul Kalam Azad*	114340	1974
	35. Maharshi Karve*	123450	1978
2. Great Eastern	36. Jag Laadki*	103124	1972
Shipping Co.	37. Jag Laxmi*	121423	1976
(5)	38. Jag Leela*	125085	1976
	39. Jag Jyoti	17892	1968
	40. Jag Prakash	25480	1974
	41. Jag Priya	32878	1975
3. Varun	42. Arya Doot	20968	1960
Shipping Co. Ltd. (2)	43. Meghdoot	6102	1973
4. Essar Bulk Carries Ltd. (1)	44. Nand Kala	6703	1972
5. Sagar Line (India) Pvt. Ltd. (1)	45. Sugar Tarni	712	1965
6. Scindia Steam Nav. Co. Ltd. (1)	46. Walchand*	104750	1973
7. Ratnakar Shipping Co. (1)	47. Ratna Jayashree	60252	1966

*Ore/Oil Bulk Carriers
Source : Ministry of Shipping & Transport

(iv) The oil Coordination Committee synchronizes the movement of coastal tankers from port to port. The task of the oil coordination committee is to ensure both the optimal utilization of the available fleet and the product movement requirements. Monthly coastal plans are prepared for all the products containing details about Loadport, disport, Laydays and Tankers. However, the distribution plan is not based on a precise measure of the cost elements involved in moving the product from port to port by tankers. It is rather a judgemental approach wherein the surplus ports move the products to the nearby deficit ports. Hence it is of paramount importance, that due recognition is given to the cost elements and port traffic arrived at. The Report of the National Transport Policy Committee observed that in planning port development, a certain degree of foresight and advance planning is essential because construction of port capacity invariably entails long gestation lags. To this end it is useful to have accurate and realistic estimates of port traffic as huge amounts of capital investment are involved in port development. (Planning Commission, Government of India, 1980).

5.3 The Approach in this Study : Problem Parameters

(i) *Supply and demand estimates* : The refinerywise achieveable crude intake and crude mix as given in OCC's long term production estimates have been adopted. Average monthly demand at ports were furnished by OCC for all the products. The year considered is 1983-84. Tables 5.III to 5.IX contain the average monthly demand and supply of the ports for the products.

(ii) *Cost of tanker transportation* : Many revisions of the method of computing average tanker transportation costs or rates have been necessary in the past because data were scarce. Any scrap of information tended to call for a revision (Nelson, W.L.; 1968). Most of the readily available information about costs are the spot or single charter rates published by some brokers such as Mullion Tanker (Shipbrokers) Ltd. (Nelson, W.L.; 1976).

(a) World scale rates—World scale rates are a copy-righted method of describing and comparing tanker rates. It is sponsored by the Association of Shipbrokers and Agents, U.S.A. and the International Tanker Freight Ltd., London.

(b) Spot rates—A producer or company may need to ship a single cargo. He applies to a shipbroker familiar with the availability of vessels of various sizes. A spot or single charter rate is negotiated and stated in world scale. If many vessels are available the rate will be low and vice versa. (Nelson, W.L.; 1976).

(c) Nelson's average tanker transportation cost : Nelson's

average tanker transportation costs attempt to account for the costs of (a) shipments in company owned (or government) vessels by means of the break even cost (plus profit) of operation, (b) Long-term charters (rented ships) by means of the 36-month running average of single charter rates, and (c) the few spot shipments by means of the 12-month average of single charter rates. These components are related in Nelson's equation for both short and long voyages. (Nelson, W.L.; 1968).

(d) AFRA—The results of the average freight assessment of AFRA as there are commonly known, are calculated by the London Tanker Brokers' Panel and are published on the first business day of each month. Results are provided for five different deadweight groups. The assessments represent in terms of world scale the weighted average cost of commercially chartered tonnage as employed in the international transport of oil. AFRA rates are sometimes the agreed basis adopted by oil companies and governments in determining average tanker transportation costs. (Bevan, E.A.; 1976).

(e) In India, the role of coastal shipping of bulk commodities such as coal, cement and oil products has been considered by the Committee on Transport Policy and Coordination (1966). The World Bank Coal Transport study team (1964) has worked out the 'Cost to the economy' of rail cum sea movement. The economic cost of coastal shipping has not been estimated by the Joint Technical Group on Transport Planning in their studies (1966, 1967).

(iii) In the absence of any readily available information regarding cost for movement of products from port to port, an attempt is made in this study to arrive at the cost per deadweight tons (dwt). Deadweight ton is the total carrying capacity of a ship expressed in long tons (2240 pounds). It is the difference between a ship's displacement when fully loaded and its displacement with no cargo, fuel, ballast, stores, passengers or crew but with all machinery "wet" and ready for steaming.

The cost per deadweight ton is arrived at for a tanker of 25,000 dwt capacity. The cost elements considered are victualling cost (i.e. hire charges of the tanker) and fuel consumption during transit. However, port charges and fuel consumption in the port are not considered. Distances from port to port in terms of nautical miles are given in Table 5.X. Average speed of a tanker of 25,000 dwt capacity has been assumed as 13.5 nautical miles per hour. The distances in terms of nautical miles have been converted into voyage time expressed in hours with the speed of 13.5 nautical miles per hour (Table 5.XI). The voyage time is expressed in days as summarised in Table 5.XII. The details of the cost elements are given in Table 5.XIII, and the cost per dwt per day has been arrived at. The product of the voyage time in days (Table 5.XII) and the cost per dwt per day gives the cost per dwt from port to port as summarised in Table 5.XIV.

5.5 The Model

A number of industries are concerned with the distribution of a group of products using the same facilities. In many instances it is possible to consider this type of problem as an extension of the normal transportation problem and to develop a special computation routine for its solution (Haley, 1966).

'p' different refinery products are available at 'm' loadports. The products are distributed to 'n' disports at a cost which depends on the locations of 'm' and 'n' ports.

Denoting the amount that is sent from loadport 'i' to disport 'j' of the refinery product 'k' by x_{ijk} and the cost of sending one unit by C_{ijk} then it is required to

Minimize

$$\sum_{i=1}^{m} \sum_{j=1}^{n} \sum_{k=1}^{p} X_{ijk} C_{ijk}$$

Subject to

Supply constraints of the loadports

$$\sum_{j=1}^{n} X_{ijk} = A_{ik}$$

$$(i=1,2,\ldots, m) \quad (k=1,2,\ldots, p)$$

Demand constraints of the disports

$$\sum_{i=1}^{m} X_{ijk} = B_{jk}$$

$$(j=1,2,\ldots, n) \quad (k=1,2,\ldots, p)$$

$$X_{ijk} \geq 0, \text{ for all i, j and k}$$

TABLE 5.III : **Product : Motor Spirit**

Light Distillate :

Adjusted Average Monthly Production and Demand

(*Figures in '000' MT*)

	Haldia	Kandla	Okha	Bombay	Goa	Cochin	Mangalore	Madras	Vizag	Supply
Bombay										27
Cochin										14.9
Madras										7.9
Vizag										8.9
Haldia										12.0
Demand	11.4	0.3	0.4	26.4	1.9	13.7	1.2	9.0	4.9	

Source : Oil Coordination Committee's long term production estimates.

TABLE 5.IV : **Product :—Naptha**

Adjusted Average Monthly Production and Demand

(*Figures in '000' MT*)

	Bombay	Cochin	Madras	Vizag	Haldia	Tuticorin	Mangalore	Goa	Supply
Bombay									77.5
Cochin									24.4
Madras									18.3
Vizag									12.9
Haldia									17.1
Kandla (Surplus from Koyali & Mathura Refineries)									60
Demand	34.1	18.3	18.3	6.2	—	19.2	13.3	16.3	

Source : Oil Coordination Committee's long term production estimates.

Middle Distillates

TABLE 5.V : **Aviation Turbine Fuel**

Adjusted Average Monthly Production and Demand

(*Figures in '000' MT*)

	Haldia	Okha	Bombay	Goa	Cochin	Mangalore	Madras	Vizag	Supply
Bombay									34
Cochin									1.9
Madras									12.4
Haldia									9.4
Demand	7.7	1.8	37.3	0.4	1.7	0.2	8.7	0.7	

Source : Oil Coordination Committee's long term production estimates

TABLE 5.VI : **Light Diesel Oil**

Adjusted Average Monthly Production and Demand

(Figures in '000 MT)

	Haldia	Kandla	Okha	Bombay	Goa	Cochin	Mangalore	Madras	Vizag	Supply
Bombay										31.4
Cochin										2.1
Madras										4.8
Vizag										3.5
Haldia										20
Demand	31.2	0.8	1.8	20.8	0.7	1.8	0.3	4.8	3.5	

Source : Oil Coordination Committee's long term production estimates.

TABLE 5.VII : IIHigh Speed Diesel

Adjusted Average Monthly Production and demand

(Figures in '000' MT)

	Haldia	Kandla	Okha	Bombay	Goa	Cochin	Mangalore	Madras	Vizag	Supply
Bombay										256.6
Cochin										86.2
Madras										46,7
Vizag										51.6
Haldia										49.5
Demand	19.3	3.3	3.4	149.7	14.2	84.6	6.4	106.1	70.2	

Source : Oil Coordination Committee's long term production estimates.

TABLE 5.VIII : **Superior Kerosine Oil**

Adjusted Average Monthly Production and Demand

(Figures in '000' MT)

	Haldia	Kandla	Okha	Bombay	Goa	Cochin	Mangalore	Madras	Vizag	Supply
Bombay										24
Cochin										34
Madras										33.5
Vizag										2.2
Haldia										15.0
Demand	38.9	1.2	2.1	98.1	5.2	31.1	2.3	51.7	25.3	

Source : Oil Coordination Committee's long term production estimates.

TABLE 5.IX : **Furnace Oil**

Heavy Distillate

Adjusted Average Monthly Production and Demand

(*Figures in '000' MT*)

	Haldia	Kandla	Okha	Bombay	Goa	Cochin	Mangalore	Tuticorin	Madras	Vizag	Supply
Bombay											21.8
Cochin											14.7
Madras											32.7
Haldia											33.7
Demand	72.8	1.9	2.3	103.9	17.9	14.7	2.0	1.5	19.6	20.0	

Source : Oil Coordination Committee's long term production estimates.

TABLE 5.X : **Distance Matrix**

(Figures in Nautical Miles)

From / To	Kandla	Okha	Bombay	Mangalore	Goa	Cochin	Tuticorin	Madras	Vizag	Haldia
Kandla	0	79	415	793	602	953	1178	1834	2078	2427
Okha	79	0	346	726	533	884	1109	1755	2019	2358
Bombay	415	346	0	425	232	585	810	1458	1322	2061
Mangalore	793	726	425	0	193	160	385	1045	1300	1639
Goa	602	533	232	193	0	365	590	1238	1502	1841
Cochin	953	884	585	160	365	0	225	876	1140	1479
Tuticorin	1178	1109	810	385	590	225	0	730	1006	1338
Madras	1834	1755	1458	1045	1238	876	730	0	328	724
Vizag	2088	2019	1322	1300	1502	1140	1006	328	0	391
Haldia	2427	2358	2061	1639	1841	1479	1338	724	391	0

Source : Oil Coordination Committee's long term production estimates.

TABLE 5.XI : Voyage Time (Assuming an Average Speed of 13.5 Nautical Miles Per Hour)

(Figures in Hours)

From / To	Kandla	Okha	Bombay	Mangalore	Goa	Cochin	Tuticorin	Madras	Vizag	Haldia
Kandla	0	5.85	30.74	58.74	44.59	70.59	87.25	135.85	154.66	179.77
Okha	5.85	0	25.62	53.77	39.48	65.48	82.14	130	149.55	174.66
Bombay	30.74	25.62	0	31.48	17.18	43.33	60	108	97.92	152.66
Mangalore	58.74	53.77	31.48	0	14.29	11.85	28.51	77.40	96.29	121.40
Goa	44.59	39.48	17.18	14.29	0	27.03	43.70	91.70	111.25	136.37
Cochin	70.59	65.48	43.33	11.85	27.03	0	16.66	64.88	84.44	109.55
Tuticorin	87.25	82.14	60	28.51	43.70	16.66	0	54.07	74.51	99.11
Madras	135.85	130	108	77.40	91.70	64.88	54.07	0	24.29	53.62
Vizag	154.66	149.55	97.92	96.29	111.25	84.44	74.51	24.29	0	28.96
Haldia	179.77	174.66	152.66	121.40	136.67	109.55	99.11	53.62	28.96	0

TABLE 5.XII : Voyage Time

(Figures in Days)

From / To	Kandla	Okha	Bombay	Mangalore	Goa	Cochin	Tuticorin	Madras	Vizag	Haldia
Kandla	0	0.243	1.28	2.44	1.85	2.94	3.63	5.66	6.44	7.49
Okha	0.243	0	1.06	2 24	1.645	2.72	3.42	5.41	6.23	7.27
Bombay	1.28	1.06	0	1.31	0.715	1.80	2.5	4.5	4.08	6.36
Mangalore	2.44	2.24	1.31	0	0.595	0.493	1.187	3.225	4.012	5.05
Goa	1.85	1.645	0.715	0.595	0.	1.126	1.82	3.82	4.63	5.68
Cochin	2.94	2.72	1.80	0.493	1.126	0	0.694	2.70	3.518	4.564
Tuticorin	3.63	3.42	2.5	1.187	1.82	0.694	0	2.25	3.104	4.129
Madras	5.66	5.41	4.5	3.225	3.82	2.70	2.25	0	1.012	2.23
Vizag	6.44	6.23	4.08	4.012	4.63	3.518	3.104	1.012	0	1.20
Haldia	7.49	7.27	6.36	5.05	5.68	4.564	4.129	2.23	1.20	0

TABLE 5.XIII : Cost details for a 25,000 dwt tanker (excluding port charges and fuel consumed in the Port)

(a) Victualling Cost (occ estimate)	— Rs 100 per dwt per month	= Rs. 3.33 per dwt per day
(b) Fuel charges (occ estimate)	—	
(1) Furnace oil	— 35 MT of furnace oil per day @ Rs. 1690 per tonne	= Rs. 59,150 per day
(2) Diesel	— 2.5 MT of diesel oil per day @ Rs. 3400 per tonne	= Rs. 8,500 per day
		Rs 67,650 per day
	or	= Rs. 2.706 per dwt per day
	TOTAL of (a)+(b)	= Rs. 6.036 per dwt per day

TABLE 5.XIV : Cost per Deadweight Ton (dwt) (Excluding Port Charges & Fuel Consumed in the Port)

(*Figures in Rupees*)

From / To	Kandla	Okha	Bombay	Mangalore	Goa	Cochin	Tuticorin	Madras	Vizag	Haldia
Bombay	7.73	6.40	0	7.91	4.32	10.9	15.09	27.16	24.63	38.39
Cochin	17.75	16.42	10.86	2.98	6.80	0	4.19	16.30	21.23	27.54
Madras	34.16	32.65	27.16	19.47	23.06	16.30	13.58	0	6.11	13.46
Vizag	38.87	37.60	24.63	24.22	27.95	21.23	18.74	6.11	0	7.24
Haldia	45.21	43.88	38.39	30.48	34.28	27.55	24.92	13.46	7.24	0

The 'n' loadports are Bombay, Cochin, Madras, Haldia, Vizag and Kandla.

The 'm' disports are Kandla, Okha, Goa, Mangalore, Tuticorin, Bombay, Madras, Vizag and Haldia.

The 'p' refinery products are

(1) Light distillate—Motor spirit, Naphtha

(2) Middle distillate—Avaiation Turbine Fuel,
Superior Kerosene oil,
High Speed Diesel,
Light Diesel oil

(3) Heavy distillate—Furnace Oil

C_{ijk} is the cost per deadweight ton (dwt) of the kth product distributed from the ith loadport to the jth disport (Table 5.XII).

This problem is very similar to the transportation problem, excepting the fact that there are three suffixes instead of two.

5.5 Suggested Productwise Monthly Distribution Plan for the year 1983-84

The productwise monthly distribution plan as obtained from the solution of the model is as follows :

(i) Light Distillates

(a) Motor Spirit (M.S.)

Loadport	Disport	Quantity ('000' MT)
Bombay	Bombay	26.4
Bombay	Kandla	0.2
Bombay	Okha	00.4
Cochin	Cochin	13.7
Cochin	Mangalore	1.2
Madras	Madras	7.9
Vizag	Vizag	4.9
Vizag	Kandla	0.1
Vizag	Goa	1 9
Vizag	Madras	1.1
Haldia	Haldia	11.4

There will be a surplus of 600 MT in Haldia and 900 MT in Vizag, after the requirements at all the ports are met. The problem of Motor spirit surplus disposal would be further accentuated with the Commissioning of additional secondary processing facilities at several refineries. The Oil Coordination Committee may examine the export outlets from Haldia and Vizag ports.

(b) Naphtha

Loadport	Disport	Quantity ('000' MT)
Bombay	Bombay	34.1
Bombay	Goa	16.3
Bombay	Mangalore	13.3
Bombay	Tuticorin	13.1
Cochin	Cochin	18.3
Cochin	Tuticorin	6.1
Madras	Madras	18.3
Vizag	Vizag	6.2

There will be a surplus of 690 MT in Bombay, 17100 MT in Haldia, and 6700 MT in Vizag. Surplus of 60000 MT in Kandla, received from other refineries is fully for export.

(*ii*) *Middle Distillates*

(a) Aviation Turbine Fuel (ATF)

Loadport	Disport	Quantity ('000' MT)
Bombay	Bombay	34.0
Cochin	Cochin	1.7
Cochin	Bombay	0.2
Madras	Madras	8.7
Madras	Okha	1.0
Madras	Goa	0.4
Madras	Bombay	2.3
Haldia	Haldia	7.7
Haldia	Mangalore	0.2
Haldia	Bombay	0.8
Haldia	Vizag	0.7

The shortfall in Okha port of 800 MT can be met from Koyali refinery. Otherwise the All India Production of ATF can be matched, so as to meet the demand, by making marginal adjustments at the refineries.

(b) Superior Kerosene Oil (SKO)

Loadport	Disport	Quantity ('000' MT)
Bombay	Bombay	24.0
Cochin	Cochin	31.9
Cochin	Mangalore	2.3
Madras	Madras	33.5
Vizag	Vizag	2.2
Haldia	Haldia	15.0

Thus SKO movement will be only from Cochin Port to Mangalore. The deficit at Bombay, Madras, Vizag and Haldia would be met ex-imports.

(c) High Speed Diesel (HSD)

Loadport	Disport	Quantity ('000' MT)
Cochin	Cochin	84.6
Cochin	Madras	1.6
Madras	Madras	16.7
Vizag	Vizag	51.6
Haldia	Haldia	49.5
Bombay	Bombay	149.7
Bombay	Kandla	3.3
Bombay	Okha	3.4
Bombay	Goa	14.2
Bombay	Mangalore	6.4
Bombay	Madras	57.79
Bombay	Vizag	18.6
Bombay	Haldia	3.2

Bombay port would continue to be surplus in HSD, the total quantum of surplus being 106900 MT which would move coastally to practically all the ports which would be deficit, excepting Cochin. The deficit in Haldia port would be about 38600 MT, which would have to be met by imports.

(d) Light Diesel Oil (LDO)

Loadport	Disport	Quantity ('000' MT)
Bombay	Bombay	20.8
Bombay	Kandla	0.8
Bombay	Okha	1.8
Bombay	Goa	0.7
Bombay	Mangalore	0.3
Bombay	Haldia	7.0
Cochin	Cochin	1.8
Cochin	Haldia	0.3
Madras	Madras	4.8
Vizag	Vizag	3.5
Haldia	Haldia	20.0

The marginal deficit of 3900 MT at Haldia would have to be met from the refineries in the Eastern Sector.

(iii) Heavy Distillate

(a) Furnace Oil

Loadport	Disport	Quantity ('000' MT)
Bombay	Bombay	21.8
Cochin	Cochin	14.7
Madras	Madras	19.6
Madras	Vizag	13.1
Haldia	Haldia	33.7

All the ports, excepting Madras will be deficient in Furnace Oil. Thus, furnace oil will have to be imported at all the ports except Madras. This is in confirmity with the government plan for 1983-84.

5.6 An Evaluation of the Existing Distribution Plan and the Suggested Productwise Distribution Plan

(i) The productwise distribution plan has been arrived at in such a manner, that the overall transportation cost from port to port will be minimized. As has been pointed out earlier, the existing distribution plan is not based on a precise measure of the cost elements involved in moving the petroleum products from port to port by tankers. Though the confluence of the suggested plans with the oil coordination committee plans are evident in many respects, there are some points of differences. The Coastal movement of High Speed Diesel (HSD) is taken up for this purpose.

(*ii*) *Cost Comparison*

Suggested Plan (Monthly)

Loadport	Disport	Quantity ('000' MT)
Cochin	Madras	1.6
Bombay	Kandla	3.3
Bombay	Goa	14.2
Bombay	Mangalore	6.4
Bombay	Madras	57.79
Bombay	Vizag	18.6
Bombay	Madras	3.2

Existing Plan (Monthly) Year 1983-84

Loadport	Disport	Quantity ('000' MT)
Cochin	Mangalore	1.6
Bombay	Haldia	41.8
Bombay	Kandla	3.3
Bombay	Okha	3.4
Bombay	Goa	14.2
Bombay	Mangalore	4.8
Bombay	Vizag	18.6
Bombay	Madras	20.8

Taking the cost arrived at in (Table 5.XIV) as the basis. The total monthly cost for the suggested plan is Rs. 21,12,680. For the existing plan the total cost is Rs. 25,13,300. Hence the net saving in cost for a month is Rs. 4,00,620.

5.7 Conclusion

(i) The distribution pattern arrived at in this study, for light, middle and heavy distillate products provides a basis for accurate and reliable traffic estimates at the ports in the country. Two considerations are of paramount relevance. First, as construction of port capacity invariably entails a long gestation lag due to technical and other procedural reasons, planning for port capacity expansion must be undertaken long before congestion builds up at ports. Secondly, since port capacity can only be used at specific locations and is not transferrable to other locations without involving huge losses, there is a great investment risk, if irrational traffic projections are made.

(ii) It will be useful to relate port traffic projections to existing port capacities to get an idea of the facilities required for the port development in the next two decades or so. A proper assessment of port capacity is essential both for optimal utilisation of existing capacity and addition to it, to meet growing distribution needs for the petroleum industry. A number of attempts have been made to measure port capacity in the past, but there appears to be no consensus on the methodology used or estimates made for it. (Report of the National Transport Policy Committee; Planning Commission; Government of India, 1980). For an efficient tanker terminal the concept should be "A jetty should wait for ship and not the ship for jetty". Therefore in this respect it should be recognised that capacity of the port to handle traffic should be much in excess of projected through put.

(iii) As regards berth occupancy theoretically a berth can be used for all the 365 days of a year. In actual practice, however, berth occupancy is limited by the number of days a berth must be left free for maintenance, dredging and repair work. The working group of ports for the 6th five year plan had envisaged 75% berth occupancy as adequate for the oil harbours. Experience has shown that at the level of 75%, the ports are heavily congested because of non-availability of berth. The berth occupancy ratio internationally considered optimal is 67 percent. Berth occupancy of over 75 percent leads to poor port maintenance and hence to deterioration in the quality of port services. As against these norms the present berth occupancy at Bombay port, for instance is estimated at about 90 percent, which shows the pressure of demand at the port. The results of the present study indicate immense pressure on both Bombay and Cochin ports. Bombay port is to distribute the maximum quantity of High Speed Diesel as compared to other ports. Also Cochin is to handle seven major products. The utilisation of berths on 90 percent scale is considered extremely damaging to efficient operation of the port system. In the interest of efficient port operation, it is absolutely essential that the high level of berth occupancy is brought down.

(iv) The time taken by the tanker on the berth, that is the berth occupancy depends upon the performance of tanker with respect to its pumping capacity. At present tankers average rate of discharge for white oil is 400 MT per hour. For discharge of a parcel of 25,000 MT of a single product a tanker needs 62 hours time. Depending on various factors, apart from her loading/discharging activity, a tanker takes approximately 14 to 15 hours on the berth for other operations as per oil coordination committee's estimate.

(a) Berthing time	2 hours
(b) Connecting house/ullage/Cargo assessment	1½ hours
(c) Empty tank inspection	1 hour
(d) Ballast	6 hours
(e) Hose disconnection	1 hour
(f) Documentation & waiting for Pilot	3 hours
	14½ hours

Allowing for 14½ hours required for the other activities, a tanker loaded with a single product takes approximately 74½ hours on berth. This is abnormally high when compared to accept norms of 36 hours permitted as per world scale.

(v) It is economical to transport petroleum Cargo in bigger vessels. Studies show a possible reduction of 20% to 30% in transportation cost between a 30,000 dwt and a 70,000 dwt tanker (Per Bruun; 1976); (Heggmsnes, N.; 1974). Limitations of drafts in Indian ports either have resulted in dead freighting of tankers or have led to deployment of smaller vessels, consequently pushing up per ton transportation cost. Considering this, the drafts, in Indian ports need to be improved by capital dredging and maintained at the optimum level. This requires augmentation of the existing dredging capacity by the Dredging Corporation of India Ltd. and the respective ports. Also huge capital investment is needed for this purpose. To determine the transportation costs savings due to channel deepening it is necessary to consider (a) the anticipated traffic, (b) investment in the project covering the cost of tankers and channel development, (c) channel depths in the river estuary to permit safe navigation and (d) the dredging quantity required to obtain the necessary channel depths. The Oil Coordination Committee has recommended that the drafts at Kandla need to be improved to 39′, at Bombay to 42′, at Madras to 57′, at Vizag to 55′ and at Haldia to 40′. However, such decisions should be arrived at in a scientific manner. A simulation model applied the methodology to the subsidiary dock system at Haldia, 57 nautical miles seaward of Calcutta (Bhattacharya, S.K. and A.N. Biswas; 1973).

(vi) Port capacity is a dynamic term, which is based upon the tanker capacities and the shore facilities. Tanker pumping capacity normally conforms to the size of the vessel and has kept pace with the developing tanker technology. But the shore facilities in the ports have largely been responsible for restricting tanker discharge rates. As most oil terminals are located at a distance of 10-15 kms. from the berths, tanker pumps are unable to discharge cargoes at such distances. To have improved turnaround, tanker discharge needs to be boosted up by providing boosters of adequate capacity in between tanker and the storage tanks.

(a) Conventional system of using hoses for tanker discharge restricts the discharge pressure and the rate. It has been estimated by the Oil Coordination Committee that in a port like Bombay by fitting loading arm the direct saving for a single berth in terms of tanker time will work out to 200 hours for one year.

(b) In case of limited docklines, when more than one product is discharged through such docklines, the tanker pumps are used to displace and flush the lines of the previous cargo. This results in considerable delay to tankers. By equipping independent flushing pumps on berth, the turnaround time of tanker will be reduced.

(c) Thus mechanical loading arms, boosters and flushing pumps can improve considerably the turnaround period of a tanker.

(vii) While in most ports the docklines are owned by the oil industry, in ports of Bombay and Kandla the docklines are owned by the port authorities. These docklines have aged over the years, resulting in pressure restrictions and low discharge rates. Additional docklines both at Bombay and Kandla are to be provided to remove the bottlenecks.

(viii) Though the ports have been gradually replacing their older flotilla/logistic crafts, they have not been able to have additional tugs, mooring boots etc. Adequate additional tugs in the major ports of Bombay, Calcutta, Madras and Cochin need to be acquired to avoid unnecessary detention to tankers.

(ix) In most Indian ports navigation is restricted to day light hours only. Tankers having completed discharge or loading have to wait for the day light hours, to move. This results in as much as 12 hours detention in most cases. In some cases, where the ports are tidal, the situation is even worse. Hence necessary arrangements should be made to make the ports operational round the clock.

(x) The existing communication network within the port and around the coast are rather inadequate. These have to be updated to meet the latest requirements of the shipping industry Even normal messages get delayed and ships have to wait at anchorage on account of poor communication between the oil companies, ship's agents and various other concerned agencies, through coast/port radio system. Importance of proper communication between the tankers and the shore terminal has been recognised by various developed nations and their communication network has been updated to avoid unnecessary detention to vessels.

(xi) Electronic equipment has come to play a significant role in the port operations. In developing nations, radar control of ships is being extensively used for traffic controls. Apart from increasing the safety, such an arrangement will facilitate proper monitoring and faster transit of vessels.

(xii) Inadequate pilot strength in the ports often results in tanker's sailings/berthings being postponed or delayed. Positioning of additional pilots does not involve high capital expenditure. Perhaps, the reason for the ports having failed to attract suitable pilots in the above category is possibly because of high remuneration which a qualified pilot can earn through the shipping industry. A fresh look needs to be given to this aspect to induct in proper qualified and trained personnel for pilotage duties.

(xiii) Pollution of seas is a heavy price to pay for industrialisation and, therefore, protection of environment should be given proper importance. Hence all oil handling ports should be adequately equipped with only water reception facilities. These facilities at present are limited to ports of Bombay, Haldia and Vizag. Existing facilities in these ports are quite old and are inadequate and have to be updated. Necessary oily ballast docklines, oily water separators should be provided in all the oil handling ports.

Special emphasis must be given for pollution control in Cochin port, in view of the flourishing fishing industry in the area. Operational pollution caused by pumping dirty ballast into oil reception facilities can be controlled, but the accidental discharge which is a major source of pollution of seas can only be controlled by means of having proper equipment such as skimmers, oil booms, recovery boats and dispersants. Though the Indian Navy Coast Guard has been entrusted with the task of combating the pollution of seas, yet ports have to gear themselves to contain and recover oil spills within the port limits. Therefore acquisition of necessary equipment to combat pollution in the ports should be planned.

(xiv) As port capacity is directly related to the level of its labour productivity, which can be improved by better organisation and modernisation of facilities, port labour should be taken into confidence whenever schemes are mooted for port modernisation. Compared to ports elsewhere, Indian ports have a larger labour force for identical operations. Moreover, on the one hand, there is an excess of labour for most port handling operations; on the other there is an acute shortage of specialised labour such as pilots and berthing masters, affecting efficiency of port operations adversely.

(xv) Port congestion is a recurring problem at our major ports. Tankers arrive in a random fashion and congestion occurs when they arrive in a bunch. Unfortunately the arrival pattern of tankers is not within the control of port authorities. When the tankers arrive they may be able to move directly on to a berth or they may have to wait until other tankers are cleared before they can do so. The amount of Cargo which each ship carries and the time required to handle it will also vary, so that the delay to tanker waiting for berths will be the resultant of two variable functions. Operations Researchers have developed mathematical techniques for analysing queueing problems of this nature. (e.g. Mettam, J.D.; 1967). It is strongly recommended that recourse to such techniques be made, for the planning and design of Indian ports, to forecast realistically the likelihood of congestion and costly delays as tankers wait for berths. Cost of delays may be assessed, based upon actual costs or demurrage charges as appropriate. The cost of the delays to tankers waiting for berth can be used in the cost/benefit analysis for comparing the economics of developing additional facilities or of continuing to operate existing facilities with increasing congestion. In the chapter to follow, an application of queueing theory for distribution from oil depots will be considered. Nevertheless, the technique, it is hoped can be effectively employed for optimisation of port facilities too.

Glossary of Terms

(i) Ports : Ports are points where the physical equipment for the transfer of goods from water to land trasportation or vice versa and from large ocean going ships to small coastal vessels or river barges is installed.

(ii) Deadweight tons : The total carrying capacity of a ship expressed in long tons (2240 pounds) is known as dead-weight ton. A deadweight ton is roughly equivalent to 7-7.5 barrels of crude oil, depending on the specific gravity of oil.

(iii) Berth : The water area, at the waterfront edge of a wharf, reserved for a ship is known as berth.

(iv) Demurrage : A term widely used in the shipping industry to quantify the amount of (and liability for) any additional costs incurred by a loading/unloading vessel arising from delays and lost time.

(v) Draft : Depth a vessel sinks when afloat, as measured from water line to the lowest immersed part of hull is known as draft i.e. it refers to the depth required to float a vessel.

6

DISTRIBUTION OF A MIDDLE DISTILLATE PRODUCT FROM UPCOUNTRY DEPOTS

The upcountry depots form secondary distribution centres receiving supplies from one or more of the refineries and/or main port installations by rail, road and in some locations through product pipelines. The depots are spread over various locations all over the country, with each location, having one or more depots corresponding to the number of oil companies serving at the location. Oil company owned tank lorries as well as private tank lorries move the refinery products from the oil depots to the retail outlets. Inspite of the best efforts, to fill up the tank lorries as and when they arrive, the waiting time for the tank lorries at most of the depots in the country is quite high. This chapter deals with the application of queueing theory, for having a measure of the expected waiting time of the tank lorries, taking into account the pattern of their arrival and the facilities existing in the depots. Also the additional facilities required, for conformity of the waiting time of tank lorries to that of the norm fixed by the oil companies is quantified. Consequently this may enable optimal distribution of products from the upcountry depots to the retail outlets. The middle distillate product, High Speed Diesel (HSD) is chosen for the study, because the queue is more for this product as compared to other refinery products. A study of Indian Oil Corporation's, biggest terminal at Bijwason in Haryana, was carried out for this purpose.

6.1 Tank Lorry Operation

In tank lorry operations, the ideal to be aimed at is direct delivery from the depots to the storage of the customer or the retail outlet, by a fully loaded tank lorry of the largest convenient size and with a turnaround time which eliminates idling at the loading and/or the unloading points. As for the turnaround time of the tank lorries, at most of the depots in the country, long queues of tank lorries waiting for filling up the products is a common feature. The queueing problem is acute during peak season for products like High Speed Diesel (HSD) and Light Diesel Oil (LDO). There is thus considerable scope to economise on costs as far as tank lorry operations are concerned. One cannot look merely at the size of the direct savings in this regard, for savings would also be in the form of reduced fuel consumption of the most important middle distillate product, viz. HSD, whose requirements almost directly determine the refinery capacity and/or import requirements of the country. The progressive elimination of these constraints, which have an important bearing on distribution costs, would require concerted action and long range planning. The Damle Committee report on distribution system of Petroleum products (1976) emphasized the need for a systematic study of the hours during which there is heavy congestion at the depots and the optimum manner in which this can be minimised.

6.2 Basic Structure of Queueing Models

Units requiring service are generated over time by an "input source". These units enter the queueing system and join a queue. At certain point of time, a member of the queue is selected for service by some rule known as service discipline. The required service is then performed for the unit by the service mechanism, after which the unit leaves the queueing system. This process is depicted in Figure 6.I.

6.3 System Structure

In oil depots, the system structure will consist of vertical storage tanks, a pump house with motors of varying capacity, and a tank lorry filling shed with filling bases. The depot under study viz. The Indian Oil Corporation's terminal at Bijwason handles motor spirit, High Speed Diesel, superior kerosene oil and aviation turbine fuel. Eight vertical storage

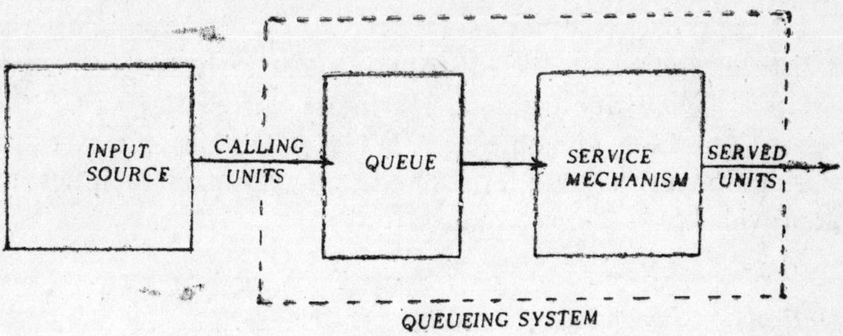

FIG. 6.I

tanks have been commissioned. Two vertical storage tanks are under construction. Products are transmitted from Mathura refinery through pipelines. The capacities of the vertical storage tanks are summarised in Table 6.I.

TABLE 6.I : **Capacities of the vertical storage tank**

Tank No.	Product	Capacity
V.S.T. No. 1	HSD	14,000 KL
V.S.T. No. 2	HSD	14,472 KL
V.S.T. No. 3	SKO	11,300 KL
V.S.T. No. 4	MS	10,300 KL
V.S.T. No. 5	MS	10,000 KL
V.S.T. No. 6	ATF	5,475 KL
V.S.T. No. 7	SKO	5,400 KL
V.S.T. No. 8	ATF	7,128 KL

The pump house has 6 motors, 2 of 50 HP and 4 of 40 HP. The motors for High Speed Diesel are of the same capacity. The tank lorry filling shed has 20 filling bases in all, out of which 10 bases are for High Speed Diesel. Indian Oil Corporation owns 23 tank lorries of various capacities, and also has 18 contractor parties to meet their requirements. Bharat Petroleum Corporation and Hindustan Petroleum Corporation have allocated 25 and 11 tank lorries respectively for this depot. Indo Burmah Petroleum Company has 14 lorries at the terminal. Besides Delhi Transport Corporation has 6 tank lorries and on an average each lorry makes 3 trips a day.

6.4 The Approach in this Study

The approach adopted involves first of all examining the statistical process of the distribution of arrivals of the tank lorries. The development examines properties of the poisson process. The queueing model quantifies the expected waiting time of the tank lorries and the additional facilities to be provided.

(i) *Problem Parameters*

(a) Mean arrival rate of tank lorries per day (λ) : For estimating the mean arrival rate of tank lorries per day, daily throughput during the working days in the months of May and June were observed as summarised in Table 6.II. Mean arrival rate of tank lorries per day is computed assuming an average of 10 kilolitre capacity per lorry and rounded off to the nearest integer, The Mean arrival rate of tank lorries per day for the depot under study is 171.

TABLE 6.II : **Mean arrival rate of tank lorries per day**

Day	Daily throughput in Kilolitres	Tank lorries
1	1766	177
2	1870	187
3	1555	156
4	2018	202
5	1406	141
6	1502	150
7	1795	180
8	1868	187
9	1729	173
10	1858	186
11	1568	157
12	1668	167
13	1738	174
14	1423	142
15	1741	174
16	1909	191
17	1513	151

18	1724	172
19	1462	146
20	1732	173
21	1448	145
22	1865	181
23	1476	148
24	1498	150
25	1564	156
26	1888	189
27	1437	144
28	1646	165
29	1725	173
30	1444	144
31	1690	169
32	1710	171
33	1800	180
34	1682	168
35	1742	174
36	1685	169
37	1597	160
38	1904	190
39	1902	190
40	2000	200
41	2004	200
42	2080	208
43	1790	179
44	1801	180
45	1709	171
46	1682	168
47	1701	170
48	1801	180
49	1607	161
50	1805	181

(b) Service rate per filling base or channel : A time study was conducted for estimating the service rate per filling base/ channel. The total time for filling up a 16 kilolitre (KL) tank lorry was 41 minutes. This time is constant, because the pumps earmarked for HSD are of the same capacity. The effective working hours per day are 7½ and 175.60 KL can be filled up. Assuming a capacity of 10 KL per lorry, 17.56 lorries can be filled up per day. Hence μ i.e. the service rate per filling base/channel is a constant at 17.56 lorries per day.

(c) Utilisation factor (ρ)

The utilisation factor ρ is computed as $\lambda/C\mu$, where

λ=Mean arrival rate of tank lorries per day

C=Number of filling bases

μ=Service rate per filling base

$\lambda=171$, $\mu=17.56$ and $C=10$

Hence $\rho=0.9738091$.

(ii) *Testing the Distribution of the Input Process*

(a) In collecting data for estimating the distribution of the number of arrivals or of holding times, the number of arrivals and the intervals of time in service are recorded as a function of time. These data may then be grouped in time intervals, after which statistical tests for independence of the means in successive time intervals may be made. With the data or portion of the data which appears to be homogeneous, a further grouping is made and a frequency histogram constructed. The appearance of this histogram suggests certain standard distributions such as the normal, poisson, or exponential, or the distribution of a sum of independently and exponentially distributed random variables. The parameters of the hypothesized distribution are then estimated from the data. The observed and model distributions are compared, using the Chi-square test.

The following analysis is carried out to test the hypothesis that the arrivals to a waiting line in the oil depots are poisson distributed with a mean arrival rate per minute equal to 0.4 (171/450) i.e. mean arrival rate of tank lorries per day divided by working time in a day expressed in minutes. The intervals for counting the number of arrivals are fixed at 2 minutes. In each such interval, the total number of arrivals is counted. Then the frequency of intervals in which a given number of arrivals occurs is counted. The arrival pattern for a particular day is shown in Table 6.III.

Tables 6.IV and 6.V show that the arrival pattern of tank lorries is described by a poisson process. Also when the times of arrival to a queue occur by a poisson process, the interarrival times have a related exponential distribution. Conversely if for example, service times have an exponential distribution, customers are admitted to service by a poisson process.

TABLE 6.III : **The arrival pattern of tank lorries a poisson process**

Time A.M.	0 Lorry	1 Lorry	2 Lorries	3 Lorries
9.00— 9.02			X	
9.02— 9.04				X
9.04— 9.06		X		
9.06— 9.08		X		
9.08— 9.10		X		
9.10— 9.12		X		
9.12— 9.14		X		
9.14— 9.16		X		
9.16— 9.18		X		
9.18— 9.20	X			
9.20— 9.22	X			
9.22— 9.24	X			
9.24— 9.26	X			
9.26— 9.28		X		
9.28— 9.30		X		
9.30— 9.32		X		
9.32— 9.34			X	
9.34— 9.36			X	
9.36— 9.38			X	
9.38— 9.40	X			
9.40— 9.42				X
9.42— 9.44		X		
9.44— 9.46		X		
9.46— 9.48		X		
9.48— 9.50	X			

Table 6.III (*Contd.*)

9.50— 9.52	X			
9.52— 9.54		X		
9.54— 9.56			X	
9.56— 9.58		X		
9.58—10.00			X	
10.00—10.02	X			
10.02—10.04	X			
10.04—10.06			X	
10.06—10.08		X		
10.08—10.10		X		
10.10—10.12		X		
10.12—10.14	X			
10.14—10.16	X			
10.16—10.18	X			
10.18—10.20	X			
10.20—10.22		X		
10.22—10.24		X		
10.24—10.26	X			
10.26—10.28	X			
10.28—10.30			X	
10.30—10.32			X	
10.32—10.34		X		
10.34—10.36		X		
10.36—10.38	X			
10.38—10.40	X			
10.40—10.42	X			
10.42—10.44	X			
10.44—10.46	X			
10.46—10.48	X			
10.48—10.50	X			
10.50—10.52	X			
10.52—10.54		X		
10.54—10.56			X	
10.56—10.58		X		
10.58—11.00 AM	X			
Total Frequency	24	24	10	2

(1) Comparison of Arrival Data with Poisson Distribution

TABLE 6.IV : **Comparsion of arrival data with poission distribution**

Number of Arrivals 'n'	Number of 2 Minutes Intervals Observed with 'n'	Value from Poisson Distribution	Number of 2 Minute Intervals Expected with 'n'
0	24	.449	26.9
1	24	.359	21.5
2	10	.144	8.6
3	2	.048	2.8

Using the Poisson distribution,

$$P_n(t) = \frac{(.4t)^n e^{-4t}}{n!}$$

with t=2, the third column is computed, for values of n from the first column. This gives the probability of the indicated arrivals. By multiplying the probabilities by the total frequency 60, we compute the number of 2 minute intervals expected to contain the indicated number of arrivals 'n'.

(2) To judge whether the poisson process is an acceptable fit statistically, Chi-square test is performed.

TABLE 6.V **Observed and Model distribution**

Observed 'O'	Expected 'E'	(O−E)	$(O-E)^2$	$(O-E)^2/E$
24	26.9	. 2.9	8.41	0.312
24	21.5	2.5	6.25	0.290
10	8.6	1.4	1.96	0.227
2	2.8	0.8	0.64	0.2228

$$X^2 = \sum \frac{(0-E)^2}{E} = 1.018$$

(v); Degrees of Freedom = K − 1 = 3

For v=3, $X^2_{0\cdot05} = 7.815$

The calculated value is much less than the table value of X^2, and hence the fit is good. Hence the arrival of tank lorries follow a poisson process, at 95% statistical level of significance.

6.5 The Model

The above discussion suggests that the arrival pattern of tan k lorries follow a Poisson process and that the service rate per channel is constant. Also there are multiple channel in the tank lorry filling shed.

Poisson input, Constant service times, multiple channels [Molina, E. C. (1922), Crommelin, C.D. (1932), Saaty, Thomas L. (1961)].

$$P(t>0)=\frac{\dfrac{(\lambda^{C}e^{-\lambda}/C!)C}{C-\lambda}}{1-e^{-\lambda}\sum\limits_{n=c}^{\infty}(\lambda^{n}/n!)+\dfrac{(\lambda^{C}e^{-\lambda}/C!)C}{C-\lambda}}$$

P(t>0)—Probability that the waiting time is greater than zero

*λ—λ/μ (Mean Arrival Rate)/(Service Rate per Channel)

e—The Base of Natural Logarithms

C—Number of Channels

$$W_q=P(t>0)\quad\frac{1}{\mu C(1-\rho)}\quad\frac{C}{C+1}\quad\frac{1-\rho^{C-1}}{1-\rho^{C}}$$

W_q—Waiting time in the queue

ρ—Utilisation factor $\frac{\lambda}{C\mu}$

μ—Service Rate Per Channel

(ii) *Application of the Model for the Existing System*

$\lambda=171$ Tank Lorries per day

$C=10$

$\mu=17.56$ Tank Lorries per day

$\rho=\frac{\lambda}{C\mu}=0.9738041$

$\lambda/\mu=9.77$

$$P(t>0)=\frac{\dfrac{(\lambda^{C}e^{-\lambda}/C!)C}{C-\lambda}}{1-e^{-\lambda}\sum\limits_{n=c}^{\infty}(\lambda^{n}/n!)+\dfrac{(\lambda^{C}e^{-\lambda}/C!)C}{C-\lambda}}$$

* To obtain a result involving μ, λ is replaced by λ/μ.

$$P(t>0)=\frac{\dfrac{(9.77^{10}\times 2.7183^{-9.77}/10!)10}{10-9.77}}{1-2.7183^{-9.77}\sum_{n=10}^{\infty}(9.77^{n}/n!)+\dfrac{(9.77^{10}\times 2.7183^{-9.77}/10!)10}{10-9.77}}$$

$$P(t>0)=0.9251$$

$$W_q=P(t>0)\ \frac{1}{\mu C(1-\rho)}\ \frac{C}{C+1}\ \frac{1-\rho^{C-1}}{1-\rho^{C}}$$

$$=0.9251\times\frac{1}{17.56\times 10(1-0.9738)}\times\frac{10}{10+1}\times\frac{1-.9738^{9}}{1-.9738^{10}}$$

$=0.1665774$ day [Consists of $7\frac{1}{2}$ Hours]

$=1.24$ Hours

$=74$ Minutes

Thus with the existing system of 10 channels [Filling Bases] for HSD, the waiting time in the queue is 74 minutes, for a tank lorry. Dimensional Analysis was carried out and the results were found to be consistent.

(iii) *Comparison of the Norm and the Waiting Time*

The norm fixed up by Indian Oil Corporation is that a 12 KL tank lorry should not spend more than 45 minutes in the system (including service time). In this study, an average of 10 KL tank lorry is assumed. Hence the norm for a 10 KL tank lorry can be taken as 37.5 minutes. The service time which is constant is 25.5 minutes. Hence the norm for waiting time (W_q) in the queue is 12 minutes (the total time spent in the system—service time). The computations carried out suggest that the expected waiting time in the queue is 74 minutes. This is also confirmed by the sample data of lorries given in the Table 6.VI. By increasing the number of channels (i.e. the filling bases) the waiting time in the queue will be reduced. Also the probability that the waiting time is greater than zero, will decrease, which is nearing unity, in the present system.

TABLE 6.VI **Comparison of the norm and the waiting time**

Lorry	Capacity KL	Time In	Time Out	Constant Service Time (minutes)	Waiting Time in the Queue (minutes)
1	10	9.05 AM	10.00 AM	26	29
2	9	9.07 AM	10.15 AM	23	45
3	12	9.20 AM	10.15 AM	31	24
4	12	9.30 AM	10.50 AM	31	49
5	10	9.35 AM	11.05 AM	26	64
6	12	9.45 AM	11.15 AM	31	59
7	12	9.45 AM	11.17 AM	31	61
8	16	9.50 AM	11.53 AM	21	82
9	12	9.50 AM	12.15 PM	31	54
10	10	9.57 AM	11.15 AM	26	52
11	10	10 00 AM	11.35 AM	26	69
12	12	10.15 AM	11.04 AM	31	54
13	12	10 30 AM	12.20 PM	31	79
14	16	10.45 AM	12.10 PM	41	44
15	10	10.47 AM	12.50 PM	26	97
16	12	12.35 PM	14.00 PM	31	54
17	12	12.30 PM	13.45 PM	31	44
18	16	12.25 PM	13.55 PM	41	49
19	9	14.05 PM	15.35 PM	23	67
20	12	14.40 PM	15.50 PM	31	39

(iv) *Calculation of $P(t)>0$, and W_q for $C=11$*

$$P(t>0)=\frac{\dfrac{(\lambda^{C}e^{-\lambda}/C!)C}{C-\lambda}}{1-e^{-\lambda}\sum_{n=c}^{\infty}(\lambda^{n}/n!)+\dfrac{(\lambda^{C}e^{-\lambda}/C!)C}{C-\lambda}}$$

$$=\frac{\dfrac{(9.77^{11}\times 2.7183^{-9.77}/11!)11}{11-9.77}}{1-2.7183^{-9.77}\sum_{n=11}^{\infty}(9.88^{n}/n!)+}$$

$$+\frac{(9.77^{11}\times 2.7183^{-9.77}/11!)11}{11.977}$$

$P(t>0)=0.6372687$

$$W_q=P(t>0)\ \frac{1}{\mu C(1-\rho)}\ \frac{C}{C+1}\ \frac{1-\rho^{C-1}}{1-\rho^{C}}$$

$$W_q=0.6372\times\frac{1}{17.56\times 11(1-.8852)}\ \frac{11}{11+1}\ \frac{1-.8852^{10}}{1-.8852^{11}}$$

$=0.025$ Day

$=0.19$ Hour

$=12$ Minutes

Hence C=11, the expected average waiting time in the queue is 12 minutes. This is the acceptable waiting time in the queue, i.e. the norm for a 10 K.L. Lorry.

6.6 Conclusion

(i)

(a) The study suggests that the filling base should be increased by one for HSD. If this is done, waiting time in the queue is twelve minutes which conforms to the fixed up norm.

(b) With the present system, the utilization factor (i.e. the ratio of mean arrival to mean service rate) is approaching unity. It is important that the traffic intensity factor remains less than unity. The input rush may create a great pressure in the system, requiring the shutting down of operation. During the months of May/June on atleast six working days, there were operational problems.

(c) If the filling base for HSD is increased by one, the utilisation factor is 0.88, and the traffic intensity will be less than that of the present system.

(d) It was observed during the study that after filling up, a tank lorry on an average waits for 5 to 10 minutes, for the operation officer's verification of the dip measurement. Waiting time on this account can be minimised, if more personnel are allocated to the tank lorry filling shed for verification of the dip measurement. One operation officer should be assigned for not more than say three or four filling bases.

(ii)

(a) It is felt that by improving the input data and by

validating all the assumptions, realistic and acceptable estimates can be developed for upcountry depots in the country.

(b) It is suggested that a technique on similar lines be developed for optimal distribution of other refinery products from upcountry depots in the country, for normal as well as peak season.

(c) A realistic norm of waiting time should be fixed for effective control in all upcountry depots. A daily report should be prepared depicting the actual waiting time and the norm and reasons for deviation should be explained for.

(d) Quick remedies for queueing inefficiencies can be brought about by reducing the service time, or the variation in the service time, by hiring additional service available for rush periods, and by controlling the arrival distribution. Smoothing out arrival irregularities which persist is also desirable. It is important that arrivals be controlled, so that the traffic intensity remains less than unity. Staggering of working hours to regularise input and use of facilities over an extended period would tend to alleviate heavy congestion at peak loads.

(e) Cost analysis are essential to many queue decisions. It is suggested that industry should carry out analysis of relevant costs, so that meaningful cost benefit analysis can be carried out. For the upcountry depot considered in this study, in the long run the benefits to be reaped by having one more filling base should far exceed the cost of filling base. This could even be quantitatively established if data regarding waiting cost due to idle time of tank lorries and service cost incorporating the capital cost of providing an additional filling base, operating and maintenance costs etc. were available. If we extend the same analogy to other refinery products as also to other upcountry depots in the country, one can well imagine the benefit to the national economy.

7

CONCLUDING FRAMEWORK

7.1

(a) The main findings and conclusions of the study of the distribution aspects from the refineries, port installations and upcountry depots have been detailed in the relevant chapters. Chapter 4 highlights the substantial financial benefits by switching over to bulk movement of bitumen from refineries. An optimal distribution plan with the aid of techniques like linear programming and goal programming can achieve tremendous reduction in costs to the economy. Also huge investments required for augmenting distribution facilities for the petroleum industry will have to be evaluated by taking recourse to sound capital budgeting appraisal techniques.

(b) The optimal distribution pattern of light, middle and heavy distillate products among the ports in the country arrived at in Chapter 5 provides a basis for accurate and reliable traffic estimates at the ports. The port traffic projections should be related to the existing port capacities. This will enable the planners to have an idea of the facilities required for the port development, so as to eliminate bottlenecks for the growing distribution needs of the petroleum industry.

(c) The appllication of queueing theory for optimal distribution of products from upcountry depots to the retail outlets as discussed in Chapter 6 can be successfully extended to refineries and ports in the country. As mentioned in Chapter 5 tanklorries in filling up products from refineries, take a lot of time due to factors other than the actual filling operation. Queueing theory can be effectively employed for determining the additional facilities at the refineries, for optimal distribution of the refinery products. Also recourse to Queueing theory and other optimization techniques should be made for the planning and design of Indian ports and to forecast realistically the likelihood of congestion and costlyd elays as tankers wait for berths in the ports.

7.2

The study amply demonstrates the applicability of operations research techniques for optimal distribution of refinery products in India. There are several other related areas which could be resolved with the help of such techniques. It is recommended that the Oil Coordination Committee apart from taking decisions based on the oil industry integrated model, should carry out research studies dealing with intricate distribution problems of petroleum industry. An attempt has been made to point out the areas which though not exhaustive, may be considered as a blue print for future studies.

(a) Port installation simulation : It would be useful to construct a model of a specific port terminal installation with a view to determine the optimum tankage, berthing facilities, pumping capacities etc. which would minimise the overall cost of ship demurrage and improve utilization of tankers.

(b) LPG facilities planning model : The basic problem to be studied will be the determination of the optimal number and locations where bottling plants may be established to meet the demand most economically in the long range.

(c) Tanker fleet optimization study : The study will assist in determining the optimum tanker fleet that the industry should deploy to carry crudes to the refineries and products from refineries to the port installations.

(d) Tankwagon utilisation study : At present the oil industry indicates to Railways only the volume (tonnes) to be loaded at each base. What is required is a joint oil industry and railway approach to exactly determine the wagons for each base by considering turnaround days of tankwagons.

(e) Optimum pipeline tributary area calculation : There is a need to evaluate the optimum tributary area of the proposed pipelines in the country and its impact on the overall transportation plan.

(f) Depot tankage evaluation :

(i) Establishing seasonality : Data will have to be collected and analysed to establish industry seasonality by area by product.

(ii) Maximum railway capacity : In conjunction with railways, the maximum tankwagon capacity to handle peak season load will have to be estimated.

(iii) Development of norms for tankage : A methodology will have to be developed to derive acceptable industry norms for seasonal and base tankage by considering various factors like tankage investment versus tank-wagon investment and single point deliveries versus multipoint deliveries.

(g) Data Collection projects : Most of the distribution studies will require considerable collection and analysis of industry data. Also considerable statistical analysis will be required to derive meaningful information. All these can be best handled by developing Computerized Management Information System (MIS). Areas which can be studied include Industry sales analysis by geogoraphical area, by season and industry depot throughput analysis.

7.3

(a) It is recommended that the results of such research studies should be quickly formalised by the Government, so that these form the basis for taking investment decisions regarding the augmenting of various facilities for distribution of refinery products. The responsibility for this formalisation may be entrusted to a standing high level Expert Group comprising the representatives of the Department of Petroleum, Railway Board, Ministry of Finance, oil companies and the Oil Coordination Committee. The involvement at high level in the suggested expert group of the concerned governmental and other agencies will make for quick decision in such matters.

(b) The Oil Coordination Committee should build a well indexed and a comprehensive library of studies, literature, published records on all the subjects, with which it is concerned. On many of these subjects there are practically no published studies and these would have to be painstakingly built up.

These should be updated on a continuing basis and should be available to the marketing oil companies, the various governmental agencies, researchers and others for facilitating decision making. In doing so, the Oil Coordination Committee should establish effective working arrangements with the Indian Institute of Petroleum and other similar bodies in and outside India.

(c) There are three clearly defined agencies responsible for planning and administering the supply and distribution system for refinery products, viz. the marketing oil companies, the Oil Coordination Committee and the Department of Petroleum. In determining the demarcation of responsibility of the individual units, Government may keep in view the necessity of retaining adequate incentives with the individual marketing oil companies to minimise their own costs, and to ensure product availability, through their distribution system throughout the country. Unless this is done in a judicious manner, there will be an inevitable tendency to pass on the blame for any failures in this regard.

It is hoped that this brief account of the task and problems of distribution conveys some idea of the magnitude and complexity of supplying the needs, both quantitatively and qualitatively of the country's insatiable demand for refinery products.

BIBLIOGRAPHY

Abraham, H. and V. Nastrand (1945) : *Asphalts and Allied Substances*, 5th Ed., Princeton, N.J.

Ackoff, R.L. and M.W. Sasieni (1968) : *Fundamentals of Operations Research*, John Wiley & Sons. New York.

Aronofsky, J. S. and A.C. Williams (1962) : The use of linear programming in oil production, *Management Science* 8, No. 4.

Balashova, T.V. *et al.* (1971) : Applications of Operations Research Methods to the Problems of Long Term Planning, *Proceedings of the VIII World Petroleum Congress.*

Barnes, R.M. (1958) : *Motion and Time Study*, 4th Ed., John Wiley & Sons, New York.

Bhatia, Ramesh K. (1976) : Technological and locational choices in petroleum and fertilizer industries : A multiprocess interregional programming model for India, *Indian Economic Review*, April, 1976.

——. (1983) : *Planning for the Petroleum and Fertilizer Industries—A Programming Model for India*, Oxford University Press, New Delhi.

Bhattacharya, S.K. and A.N. Biswas (1973) : Evaluation of harbour deepening projects, *Proceedings of the ASCE Journal of the Waterways*, Harbours and Coastal Engineering Division, Vol. 99, No. WW1.

Bevan, E.A. (1976) : 'AFRA—clarification,' *Oil and Gas Journal*, October 11, 1976.

Budhiraja, S.B. (1978) : *Management of Essential Commodities : Kerosene*, Indian Management, All India Management Association, June, 1978.

Catchpole, A.R. (1962) : The application of linear programming to integrated supply problem in the oil industry, *Operations Research Quarterly*, 13, No. 2.

Charnes, Cooper and Mellon (1954) : Model for Programming in Oil Companies, *Econometrica*.

Cheney, L.K. (1957) : Linear program planning of refinery operations, *Naval Research Logistics Quarterly*, Vol. 4, March, 1957.

Churchman, C.W. *et al.* (1957) : *Introduction to Operations Research*, Wiley, New York.

Crommelin, C.D. (1932) : Delay probability formulae when the holding times are constant, *P.O. Elect. Engrs. Journal*, Vol. 25.

"Damle Committee Report on Distribution System of Petroleum Productions (1976)", Ministry of Petroleum, Govt. of India.

Dantzig, G.B. (1956). *Formulating a Linear Programming Model*, The Rand Corporation, July 1956. U.S.A.

Dantzig, G.B. (1963) : *Linear Programming and Extension*, Princeton University Press, 1963.

Dantzig, G.B. and J.H. Ramser (1959) : Optimum routing of gasoline delivery trucks, *Proceedings of the Fifth World Petroleum Congress*, Section VIII.

Dasgupta, Biplab (1971) : *The Oil Industry in India : Some Economic Aspects*, Frank Cass & Co., London.

Dean, R. (1959) : *The Use of Linear Programming in Oil Refinery Operations*, *Process Control and Automatic*, February, 1959.

Foster, Wheeler Corporation (1958) : *Petroleum Refinery Survey for Government of India*, Vol. III, Marketing and Distribution Study.

Frankel, P.H. (1969) : *Essentials of Petroleum : A Key to Oil Economics*, Frank Cass & Co., London.

Garvin, John (1957) : Applications of linear programming in the oil industry, *Management Science*, March, 1957.

Gass, S.I. (1969) : *Linear Programming Methods and Applications*, McGraw-Hill, Tokyo.

Government of India (1969) : "Report of the Oil Prices Committee."

Government of India, Planning Commission (1974) : "Fuel Policy Committee Report."

Government of India, Planning Commission (1979) : "Report of the Working Group on Energy Policy".

Government of India, Planning Commission (1980) : "Report of the National Transport Committee".

Hadley, G. (1978) : *Linear Programming*, Addison Wiseley Publishing Company.

Hahn (1970) : *The Petrochemical Industry—Market and Economics*, McGraw-Hill.

Haley, K.B. (1962) : The solid transportation problem, *Operations Research*, 10, 4 July, 448.

——. (1963) : The Multi index problem, *Operations Research*, 11, 4 July, 368.

——. (1966) : A general method of solution for special structure linear programmes, *Operations Research Quarterly*, 17, 1 March, 83.

Heggemsnes, N. (1974) : *Port Economics*, Course Book, The Norwegian Shipping Academy.

Hillier, F.S. and G.J. Liberman (1974) : *Introduction to Operations Research*, Holden Day. San Francisco.

International Oil and Gas Educational Centre (1972) : *Petroleum Marketing and Transportation: New Ideas, Methods and Developments*, Gulf Publishing Company, Houston, Texas.

Jenkins, G. (1971) : *Oil Economists Handbook*, Applied Science Publishers Ltd., London.

Joint technical group on Transport Planning, Planning Commission, Govt. of India (1966) : "Commodity Transport Studies", Planning Commission.

Joint technical group on Transport Planning, Planning Commission, Govt. of India (1967) : "Rail and Road Transport Cost Statistics, Planning Commission."

Kahle, L.F. and A.J. Kelly (1959) : The role of sea transportation in the petroleum industry, *Proceedings of the Fifth World Petroleum Congress*, Section VIII.

Kanti Swarup *et al*. (1980) : *Operations Research*, Sultan Chand & Sons, New Delhi.

Ketharaman, S. *et al*. (1973) : Transportation, storage marketing and distribution of petroleum products, "United Nations

Interregional Seminar on Petroleum refining in developing countries, New Delhi."

Khurana, *et al.* (1973) Economic planning of refinery development, *Proceedings of United Nations Inter-regional Seminar on Petroleum Refinery*, New Delhi.

King, G.A.B. (1965) : *Tanker Practice*, The Maritime Press Limited, London.

Koeing, J.W.J. (1971) : Application of computers for refinery simulation, *Proceedings of the VI World Petroleum Congress.*

Lee, Sang, M. (1972) : *Goal Programming for Decision Analysis*, Auerbach, Philadelphia.

Lee, Sang, M. (1972) : Goal programming for Decision Analysis of Multiple Objectives, *Sloan Management Review* (Winter, 1972/73), Vol. XIV, No. 2.

Lee, Sang, M. and V. Jaaskelainen (1971) : Goal programming management math, *Industrial Engineering*, Vol. 3, No. 1.

Lee, Sang, M. and Moore, L.J. (1975) : *Introduction to Decision Science*, Petrocelli—Charter, New York.

Loomba, N.P. (1964) : *Linear Programming*, McGraw-Hill. N.Y.

Manoharan, S. (1974) : *The Oil Crisis—End of An Era*, S. Chand & Co. (P) Ltd., New Delhi.

Manne, Alan, S. (1956) : *Scheduling of Petroleum Refinery Operations*, Cambridge, Mass.

——. (1958) : A linear programming model of the U.S. Petroleum Refining industry', *Econometrica*, Vol. 26 No. 1, Jan. 1958.

——. (1966) : 'Programming data for the petroleum refining industry', *U.N. Bulletin for Industrialisation and Productivity*, No. 10.

Marks, A. (1982) : *Elements of Oil Tanker Transportation*, Pennwell Publishing Company, Tulsa, Oklahoma.

Marshall Sitting (1978) : *Petroleum Transportation and Production*, Noves Data Corporation, U.S.A.

Mettan, J.D. (1967) : *Forecasting Delays to Ships in Port*, Dock & Harbour Authority, London.

Molina, E.C. (1922) : The theory of probabilities applied to telephone truck problems, *Bell Systems Tech. Journal*, Vol. 1.

Narang, A.S. (1984): *Linear Programming and Decision Making*,

Sultan Chand & Sons, New Delhi.

National Council of Applied Economic Research (1971) : *Consumption Pattern of Selected Petroleum Products*, Asia Publishing House, New Delhi.

Nelson, W.L. (1958) : *Petroleum Refinery Engineering*, New York, McGraw-Hill.

——. (1968) : Tanker Transportation Costs', *Oil & Gas Journal*, June 24, 1968.

——. (1970) : *Guide to Refinery Operating Cost* (process costimating), Petroleum Publishing Co. Tulsa, U.S.A.

——. (1973) : "Capital and operating costs of refining in developing countries", Paper present at the *U.N. Interregional Seminar on Petroleum Refining*, New Delhi.

——. (1975) : What are costs of tanker transportation, *Oil and Gas Journal*, April 21, 1975.

——. (1976) : AFRA tanker rates—A correction, *Oil Gas Journal*, Oct. 1976.

Newby, W.J. and R.J. Dean (1960) : Planning for productivity in the oil industry, "Report of the Summer Meeting of the Institute of Petroleum", London.

Parico, J.A. (1967) : *Queueing Theory*, Prentice-Hall, N.J.

Parikh, J.K. (1981) : "Modelling energy demand for policy analysis", Planning Commission, Govt. of India.

Peat, Marwick, Mitchell & Company and Planning Commission (1966) : Manual of feasibility studies of public sector projects", Govt. of India.

Per Bruun (1976) : *Port Engineering*, Guelf Publishing Company, Book Publishing Division, Houston, Texas.

Prabhu, N.U. (1960) : Some results for the queue with Poisson arrivals, *Journal of Royal Statistical Society*, Section B, Vol. 22.

Prabhu, N.U. (1965) : *Queues and Inventories*, John Wiley and Sons. New York.

Robinson, P.J. (1956) : Practical applications of Operations Research to oil marketing, *Proceedings of American Petroleum Institute*, Section II, Marketing.

Saaty, T.L. (1961) : *Elements of Queueing Theory*, McGraw Hill. New York.

Sasieni, M.A. *et al.* (1959) : *Operations Research : Methods and Problems*, John Wiley & Sons. New York.

Schell, E. (1955) : Distribution of a product by several properties, *Proceedings of 2nd Symposium on Linear Programming*, DCS/Comptroller HQ US Air Force, Washington, D.C.

Sell, G. and H.A. Dossett (1958) : *Handbook of the Petroleum Industry*, George Newnes Limited, London.

Sellers, E.S. and L.C. Strong (1962) : The oil industry tomorrow, "Report of the Summer Meeting of the Institute of Petroleum", London.

Shelton, J.R. (1960) : Solution methods for waiting line problems, *The Journal of Industrial Engineering*, July-August 1960.

Sokolov, V. (1972) : *Petroleum*, MIR Publishers, Moscow.

Sumit Roy *et al.* (1982) : Optimum locations and capacities for LPG bottling plants in India, "Paper presented at the Operational Research Society of India Conference," Kharagpur, Dec. 1982.

Symonds, G.H. (1955) : Linear programming is helpful in solving gasoline refining and blending problems, *Proceedings of the Institute on Operations Research for Business and Industry*, University of California, April, 1955.

Tanzer, Michael (1969) : *The Political Economy of International Oil and the Underdeveloped Countries*, Temple Smith, London.

Van Horne, J.C. (1974) : *Financial Management and Policy*, Prentice Hall of India.

Wagner, H.M. (1980) : *Principles of Operations Research*, Prentice Hall of India.

Wani, J.K. (1971) : *Probability and Statistical Inference*, The Appleton Century Statistics, Series. New York.

Wellborn, A.S. and J.M. Griffith (1960) : *Petroleum Products Handbook* edited by Guthrie, V.B., McGraw-Hill. New York.

Ya-Lun Chou (1975) : *Statistical Analysis*, Holt, Rinehart and Winston.

INDEX